T0137038

Advances in Intelligent Systems and Computing

Volume 888

Series editor

Janusz Kacprzyk, Polish Academy of Sciences, Warsaw, Poland
e-mail: kacprzyk@ibspan.waw.pl

The series "Advances in Intelligent Systems and Computing" contains publications on theory, applications, and design methods of Intelligent Systems and Intelligent Computing. Virtually all disciplines such as engineering, natural sciences, computer and information science, ICT, economics, business, e-commerce, environment, healthcare, life science are covered. The list of topics spans all the areas of modern intelligent systems and computing such as: computational intelligence, soft computing including neural networks, fuzzy systems, evolutionary computing and the fusion of these paradigms, social intelligence, ambient intelligence, computational neuroscience, artificial life, virtual worlds and society, cognitive science and systems, Perception and Vision, DNA and immune based systems, self-organizing and adaptive systems, e-Learning and teaching, human-centered and human-centric computing, recommender systems, intelligent control, robotics and mechatronics including human-machine teaming, knowledge-based paradigms, learning paradigms, machine ethics, intelligent data analysis, knowledge management, intelligent agents, intelligent decision making and support, intelligent network security, trust management, interactive entertainment, Web intelligence and multimedia.

The publications within "Advances in Intelligent Systems and Computing" are primarily proceedings of important conferences, symposia and congresses. They cover significant recent developments in the field, both of a foundational and applicable character. An important characteristic feature of the series is the short publication time and world-wide distribution. This permits a rapid and broad dissemination of research results.

Advisory Board

Chairman

Nikhil R. Pal, Indian Statistical Institute, Kolkata, India
e-mail: nikhil@isical.ac.in

Members

Rafael Bello Perez, Universidad Central "Marta Abreu" de Las Villas, Santa Clara, Cuba
e-mail: rbellop@uclv.edu.cu

Emilio S. Corchado, University of Salamanca, Salamanca, Spain
e-mail: escorchado@usal.es

Hani Hagras, University of Essex, Colchester, UK
e-mail: hani@essex.ac.uk

László T. Kóczy, Széchenyi István University, Győr, Hungary
e-mail: koczy@sze.hu

Vladik Kreinovich, University of Texas at El Paso, El Paso, USA
e-mail: vladik@utep.edu

Chin-Teng Lin, National Chiao Tung University, Hsinchu, Taiwan
e-mail: ctlin@mail.nctu.edu.tw

Jie Lu, University of Technology, Sydney, Australia
e-mail: Jie.Lu@uts.edu.au

Patricia Melin, Tijuana Institute of Technology, Tijuana, Mexico
e-mail: epmelin@hafsamx.org

Nadia Nedjah, State University of Rio de Janeiro, Rio de Janeiro, Brazil
e-mail: nadia@eng.uerj.br

Ngoc Thanh Nguyen, Wroclaw University of Technology, Wroclaw, Poland
e-mail: Ngoc-Thanh.Nguyen@pwr.edu.pl

Jun Wang, The Chinese University of Hong Kong, Shatin, Hong Kong
e-mail: jwang@mae.cuhk.edu.hk

More information about this series at http://www.springer.com/series/11156

Saiful Omar · Wida Susanty Haji Suhaili
Somnuk Phon-Amnuaisuk
Editors

Computational Intelligence in Information Systems

Proceedings of the Computational Intelligence in Information Systems Conference (CIIS 2018)

 Springer

Editors
Saiful Omar
School of Computing and Informatics
Universiti Teknologi Brunei
Gadong, Brunei Darussalam

Somnuk Phon-Amnuaisuk
Centre for Innovative Engineering
Universiti Teknologi Brunei
Gadong, Brunei Darussalam

Wida Susanty Haji Suhaili
School of Computing and Informatics
Universiti Teknologi Brunei
Gadong, Brunei Darussalam

ISSN 2194-5357   ISSN 2194-5365 (electronic)
Advances in Intelligent Systems and Computing
ISBN 978-3-030-03301-9   ISBN 978-3-030-03302-6 (eBook)
https://doi.org/10.1007/978-3-030-03302-6

Library of Congress Control Number: 2018959234

This Springer imprint is published by the registered company Springer Nature Switzerland AG
The registered company address is: Gewerbestrasse 11, 6330 Cham, Switzerland

Preface

On behalf of the organizing committee, it is an honor and a great pleasure to welcome all of you to Brunei and to the Computational Intelligence in Information Systems (CIIS 2018) Conference.

CIIS Conference is initiated from the INNS-CIIS 2014, with help from the International Neural Network Society (INNS). CIIS Conference aims to bring together researchers from countries in the Asian Pacific Rim to exchange ideas, present recent results, and discuss possible collaborations in general areas related to computational intelligence and their applications in various domains.

This year, the international program committee constitutes 63 researchers from 15 different countries. CIIS 2018 has attracted a total of 41 submissions by authors from 14 different countries. These submissions underwent a rigorous double-blind peer review process. Of those 41 submissions, 20 submissions (49%) have been selected to be included in this book.

First and foremost, we would like to thank the keynote speakers, the invited speaker, and all the authors who have spent the time and effort to contribute significantly to this event. We would like to thank members of the technical committee who have provided their expert evaluation of the submitted papers; Arun Kumar, Springer Nature; members of the local organizing committee; members of the steering committee for their useful advice; and last but not least, Hjh Zohrah Binti Haji Sulaiman, our Vice-Chancellor.

We would also like to acknowledge the following organizations: Universiti Teknologi Brunei for its institutional and financial support and for providing premises and administrative assistance; Brunei Shell Petroleum, Bank Islam Brunei Darussalam (BIBD) and SprintVille Technologies for the generous financial support; International Neural Network Society (INNS) Brunei Regional Chapter and Springer for their technical support.

Finally, we thank all the participants of CIIS 2018 and hope that you will continue to support us in the future.

September 2018   Saiful Omar
                                                        Wida Susanty Haji Suhaili
                                                        Somnuk Phon-Amnuaisuk

Organization

Honorary Chair/Advisor

Hjh Zohrah binti Haji Sulaiman   Universiti Teknologi Brunei, Brunei
(Vice-chancellor)

Steering Committee

Chairperson

Zuruzi Abu Samah   Universiti Teknologi Brunei, Brunei
(Assistant Vice-chancellor (Research))

Members

Noor Maya binti Haji Md. Salleh (Assistant Vice-chancellor (Academic))	Universiti Teknologi Brunei, Brunei
Awang Hj Ady Syarmin bin Haji Md. Taib (Assistant Vice-chancellor (Industry and Services))	Universiti Teknologi Brunei, Brunei
Awang Chui-Hua Lim (Registrar and Secretary)	Universiti Teknologi Brunei, Brunei
Awang Hamdani bin Haji Ibrahim (Finance)	Universiti Teknologi Brunei, Brunei
Thien-Wan Au (Dean)	School of Computing and Informatics, Universiti Teknologi Brunei, Brunei

Mohamad Saiful bin Haji Omar   Universiti Teknologi Brunei, Brunei
   (Chairman CIIS 2018)

International Advisory Board

Chris Phillips   University of Tokyo, Japan
Dusit Niyato   Nanyang Technological University,
                                               Singapore
Francis Chin   University of Hong Kong, China
Ian Ruthven   Strathclyde University, Glasgow, UK
Irwin King   Chinese University of Hong Kong,
                                               China
Laszlo T. Koczy   Budapest University of Technology
                                               and Economics, Hungary
Md Shamim Ahsan   Khulna University, Bangladesh
Soo-Young Lee   Korea Advanced Institute of Science
                                               and Technology, Korea

Working Committees

Chairman and Co-chairs

Mohd Saiful Haji Omar   Universiti Teknologi Brunei, Brunei
Wida Susanty Haji Suhaili   Universiti Teknologi Brunei, Brunei

Program Chairs

Somnuk Phon-Anuaisuk   Universiti Teknologi Brunei, Brunei
Thien-Wan Au   Universiti Teknologi Brunei, Brunei

Secretariat

Haji Rudy Erwan Haji Ramlie   Universiti Teknologi Brunei, Brunei
Haji Irwan Mashadi Haji Mashud   Universiti Teknologi Brunei, Brunei

Poster Competition Chair

Serina Hj Mohd Ali   Universiti Teknologi Brunei, Brunei

Finance Chair

Hamdani bin Hj Ibrahim   Universiti Teknologi Brunei, Brunei

Logistics Chair

Muhd Robin Yong Bin Abdullah   Universiti Teknologi Brunei, Brunei

Ceremony Chair

Hj Idham Maswadi Haji Mashud   Universiti Teknologi Brunei, Brunei

Welfare and Accommodation Chair

Hj Sharul Tazrajiman Hj Tajuddin   Universiti Teknologi Brunei, Brunei

Web Master Chairs

Pg Hj Azhan Hj Pg Ahmad   Universiti Teknologi Brunei, Brunei
Muhd Safwan bin Ahman   Universiti Teknologi Brunei, Brunei
Rafidah Haji Tengah   Universiti Teknologi Brunei, Brunei

Sponsorship, Promotion and Publicity Chairs

Jennifer Nyuk-Hiong Voon   Universiti Teknologi Brunei, Brunei
Sey-Mey Yeo   Universiti Teknologi Brunei, Brunei
Siti Noorfatimah Haji Awg Safar   Universiti Teknologi Brunei, Brunei

Publishing Chair

Noor Deenina binti Hj Mohd Salleh   Universiti Teknologi Brunei, Brunei

Invitation and Protocol Chair

Ibrahim Edris   Universiti Teknologi Brunei, Brunei

Souvenir and Certificate Chair

Siti Noorfatimah Haji Awg Safar   Universiti Teknologi Brunei, Brunei

Refreshment Chair

Norhuraizah Haji Md Jaafar   Universiti Teknologi Brunei, Brunei

Car and Traffic Chair

Nurul Aqilah binti Hj Awang Jadar   Universiti Teknologi Brunei, Brunei

International Technical Committee

Abdelrahman Osman Elfaki	Tabuk University, Kingdom of Saudi Arabia
Abdollah Dehzangi	Morgan State University, USA
Adham Atyabi	University of Washington Seattle, USA
Alaelddin Fuad Yousif Mohammed	Korea Advanced Institute of Science and Technology, South Korea
Asem Kasem	Universiti Teknologi Brunei, Brunei
Atikom Ruekbutra	Mahanakorn University of Technology, Thailand
Bok-Min Goi	Universiti Tunku Abdul Rahman, Malaysia
Choo-Yee Ting	Multimedia University, Malaysia
Chuan-Kang Ting	National Chung Cheng University, Taiwan
Daphne Lai Teck Ching	Universiti Brunei Darussalam, Brunei
Ehsan Ahvar	INRIA, France
Eiad Yafi	University of Kuala Lumpur, Malaysia
Fahmi Ibrahim	Universiti Teknologi Brunei, Brunei
Florence Choong Chiao Mei	Heriot Watt University (Malaysia Campus), Malaysia
Gahangir Hossain	Texas A&M University-Kingsville, USA
Golda Brunet Rajan	GCES, India
Gyu Myoung Lee	Liverpool John Moores University, UK
Hafizal Mohamad	MIMOS, Malaysia
Hj Afzaal H. Seyal	Universiti Teknologi Brunei, Brunei
Hj Idham Maswadi bin Hj Mashud	Universiti Teknologi Brunei, Brunei
Hj Irwan Mashadi Haji Mashud	Universiti Teknologi Brunei, Brunei
Hjh Nor Zainah Siau	Universiti Teknologi Brunei, Brunei

Hj Sharul Tazrajiman Hj Tajuddin	Universiti Teknologi Brunei, Brunei
Hj Rudy Erwan Haji Ramlie	Universiti Teknologi Brunei, Brunei
Hui-Ngo Goh	Multimedia University, Malaysia
Ibrahim Edris	Universiti Teknologi Brunei, Brunei
Ibrahim Venkat	Universiti Sains Malaysia, Malaysia
Jennifer Nyuk-Hiong Voon	Universiti Teknologi Brunei, Brunei
Jessada Karnjana	National Electronics and Computer Technology Center, Thailand
Jonathan H. Chan	King Mongkut's University of Technology Thonburi, Thailand
Kabiru Rinjim	Universiti Teknologi Brunei, Brunei
Keng-Hoong Ng	Multimedia University, Malaysia
Kittichai Lavangnananda	King Mongkut's University of Technology Thonburi, Thailand
Kok-Chin Khor	Multimedia University, Malaysia
Laszlo T. Koczy	Budapest University of Technology and Economics, Hungary
Linda Sook-Ling Chua	Multimedia University, Malaysia
Lee-Kien Foo	Multimedia University, Malaysia
Md Shamim Ahsan	Khulna University, Bangladesh
Md Azam Hossain	Kyungdong University, South Korea
Md Motiar Rahman	Universiti Teknologi Brunei, Brunei
Minh S. Dao	Universiti Teknologi Brunei, Brunei
Mohamed Saleem Nazmudeen	Universiti Teknologi Brunei, Brunei
Mohammad Khairul Hasan	Korea Advanced Institute of Science and Technology, South Korea
Mohd Nazrin Md Isa	Universiti Malaysia Perlis, Malaysia
Morteza Zaker	Amirkabir University of Technology, Iran
Nazmus Shaker Nafi	Victorian Institute of Technology, Australia
Nurul Adilah Abdul Latiff	University Putra Malaysia, Malaysia
Naeimeh Delavari	Georgia Institute of Technology, USA
Palaniappan Ramaswamy	University of Kent, UK
Penousal Machado	University of Coimbra, Portuguese
Peter David Shannon	Universiti Teknologi Brunei, Brunei
Pg Hj Azhan Hj Pg Ahmad	Universiti Teknologi Brunei, Brunei
Phooi-Yee Lau	Universiti Tunku Abdul Rahman, Malaysia
Ravi Kumar Patchmuthu	Institute of Brunei Technical Education, Brunei
Sayed Mohamed Buhari	King Abdulaziz University, Kingdom of Saudi Arabia
S. H. Shah Newaz	Universiti Teknologi Brunei, Brunei

Serina Hj Mohd Ali	Universiti Teknologi Brunei, Brunei
Sey-Mey Yeo	Universiti Teknologi Brunei, Brunei
Siti Noorfatimah Haji Awg Safar	Universiti Teknologi Brunei, Brunei
Wee-Hong Ong	Universiti Brunei Darussalam, Brunei
Werasak Kurutach	Mahanakorn University of Technology, Thailand
Weng-Kin Lai	Tunku Abdul Rahman University College, Malaysia
Yun-Li Lee	Sunway University, Malaysia

Organizer

School of Computing and Informatics, Universiti Teknologi Brunei, Brunei

Technical Sponsor

Brunei Local Chapter for International Neural Network Society (INNS), Brunei

Contents

Network Centric Computing

Green Computing

Creative Computing

Intelligent Systems and Their Applications

Higher Order Spectral Analysis of Vibration Signals and Convolutional Neural Network for the Fault Diagnosis of an Induction Motor Bearings

Muhammad Sohaib and Jong-Myon Kim[(✉)]

Department of Electrical and Computer Engineering,
University of Ulsan, Ulsan, South Korea
md.sohaibdurrani@gmail.com, jmkim07@ulsan.ac.kr

Abstract. The current work is based on the need for an effective bearing fault diagnosis mechanism that can conclusively diagnose faults under variable speed conditions. Steady development has been made in the field of bearing fault diagnosis for the last couple of decades procuring satisfactory performance under steady speed scenarios. However, the performance of such algorithms deteriorates when it comes to diagnosing the bearing faults under variable speed conditions. In such conditions non-linearity and non-stationarity of the bearing signals effects the performance of the algorithms. In the proposed scheme, a higher order spectral analysis of the vibration acceleration signals is performed to obtain some characteristic peaks in two-dimensional spectral space by using bi-spectrum calculation. The calculated bi-spectra of the vibration signals are latter on provided to the convolutional neural network (CNN) to complete the desired task. The results of the proposed algorithm show satisfactory performance under variable speed conditions. The results are compared with that of two other algorithms that advocate the efficacy of the proposed model under variable shaft speed.

Keywords: Bearing fault diagnosis · Bi-spectrum analysis
Convolution neural network · Vibration signals

1 Introduction

With the enormous deployment of rotating machinery in the industry, bearing fault diagnosis is inevitable, as it is one of the essential parts of the machines to operate. The machine bearings deteriorate with the passage of time and eventually fail to work leading to machine failure [1–3]. The machine failures upshot in the huge economic losses as well as can be a threat to the safety of the people working in the facility [4, 5]. So, to evade the problem rotating machine bearing fault diagnosis is inexorable. The bearings develop cracks and spalls on the components due to harsh operating conditions in the industry. Whenever rolling element strikes the fault location on the components, it emits energy, which can be recorded in the form of signals.

© Springer Nature Switzerland AG 2019
S. Omar et al. (Eds.): CIIS 2018, AISC 888, pp. 3–12, 2019.
https://doi.org/10.1007/978-3-030-03302-6_1

Vibration acceleration signals are widely used to describe the bearing health condition as the degraded bearing signals consist of vibrations supplemented with increased level of noise. A detailed review is provided in [6, 7], where different fault diagnosis and fault-tolerant algorithms have been developed in different domains. These techniques can be customarily sorted into model-based and data-driven approaches. However, fault diagnosis in bearings has mostly been performed through data driven techniques. Data driven fault diagnosis techniques mainly consist of three major steps: (1) Signals acquisition; (2) features extraction; (3) using machine learning algorithms to classify the data into respective classes. In such techniques certain features associated with the heath state of bearing are extracted from time, frequency domain or time-frequency domain and are utilized to diagnose the faults. Moreover, spectral analysis of the signals in which defecates frequencies associated with different fault types are extracted to identify the localized faults through envelop analysis is a popular technique [8, 9]. Though, these defect frequencies or also termed as characteristic frequencies are kinematic quantities i.e., depends on shaft speed. In practical scenarios shaft speed and load angle experience random variations, which in turn make the bearing signals inherently nonstationary. Thus, nonstationary signals experience variations in the defect frequencies location in spectral domain. Fault diagnosis of bearing under variable speed by locating defect frequencies is hard and makes it inappropriate because of the random variation in the location of defect frequencies with the variations of the shaft speed. Hence in such condition these methods to be effective require tedious techniques that are hard to implement. In [10], Appana et al. suggested a convolutional neural network (CNN) based bearing fault diagnosis method. In the method a preprocessing step was introduced in the pipeline ahead of providing the signals to the CNN. The preprocessing step consisted of envelop analysis of the signals taken under low speed conditions. The resulting envelop power spectra of the signals were provided as an input to the CNN. The suggested technique resulted in better performance as compared to using raw signals as inputs to the CNN. In [11, 12], Kang et al. with the help of filter banks performed a sub-band analysis to reduce the effects of non-stationarity in the signals. In these works, with the help of mean-peaks and Gaussian mixture model-based residual component-to-defect component ratios most informative sub-band was identified and selected. The sub-band selection process was followed by feature extraction step such as relative wavelet energy and kurtosis value from the wavelet packet nodes and were used for fault diagnosis. A deep neural network (DNN) based bearing fault diagnosis scheme was developed in [13]. Sparse autoencoders were used to develop a three layered DNN. Complex envelop spectra of the input signals were calculated, so that, the frequency components associated with the faults become more vibrant and the autoencoder could extract features robustly. The proposed scheme yielded satisfactory performance dealing with shaft speed variations but the problem with the scheme is the size of the input data. A big data can result in computation complexity as well as deterioration in the performance.

The motivation behind this work includes: (1) Implementation of a fault diagnosis scheme that consists of deep learning algorithm which needs no handcrafted features; (2) to suggest a fault diagnosis model of bearings that is impermeable to the random variations in the shaft speed. The proposed fault diagnosis scheme is validated through publicly available bearing dataset [14]. A comparison of the results is given in the result section that shows the superior classification accuracy of the proposed scheme over the others.

The organization of the paper is as follows: Sect. 2 presents details about the experimental setup and information about the bearing dataset used to validate the proposed scheme. In Sect. 3 detailed description of the proposed fault diagnosis model is given. In Sect. 4 the outcomes of the proposed model are discussed, and Sect. 5 is about the conclusion of the paper.

2 Experimental Setup

In this study vibration acceleration signals acquired from rotating machine bearings are considered to check the effectiveness of the proposed model. The vibration signals were taken from a publicly available bearing data repository from Case Western Reserve University. Variable length vibrational acceleration signals recorded with the help of accelerometers from the test rig are considered in this study. The bearing data having variable shaft speed is publicly available at Case Western Reserve University (CWRU) repository [14]. A schematic representation of the test platform used during the CWRU bearing data acquisition is given in Fig. 1. It contains a 2-horsepower (hp) motor, a torque transducer, and a dynamometer. The bearings were subjected to different loads via dynamometer ranging from 0 hp–3 hp. The specifications of the bearings are provided in Table 1.

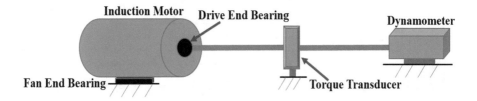

Fig. 1. A schematic diagram of Case Western Reserve University Test rig.

Table 1. Bearings specifications

Attribute	Value
Bearing model	JEM SKF 6205-2RS
Location of fault	Drive end bearings
Outer diameter	2.0472 in.
Inner diameter	0.9843 in.
Thickness	0.5906 in.
Roller diameter	0.3126 in.
Pitch diameter	1.537 in.

The motor bearings were seeded with faults (i.e., 0.007, 0.014, 0.021 in. in diameter) on the rolling elements, the inner raceways, and the outer raceways. The vibration signals were recorded at a sampling frequency of 12,000 Hz. The signals in the dataset consist of four different shaft speeds, i.e., 1722, 1748, 1772, and 1797 rounds per minute (RPMs). The variation in the shaft speed also resulted in the change of shaft load, i.e., 0, 1, 2, 3 horse powers (hp). The signals are divided into four categories, i.e., on normal and three faulty states.

3   The Proposed Fault Scheme

Figure 2 presents the proposed scheme for bearing fault diagnosis. The proposed scheme has two main steps: (1) Vibration signals segmentation into cycle length slices; (2) Fault Classification using convolution neural networks.

Vibration signals under variable shaft speed are subjected to the segmentation process. The segmentation processes resulted in multiple slices. The size of each slice is equal to the single cycle of the given signal. Since the vibration signals used for the validation of the proposed model are of variable length, so the resulted cycle length slices for each shaft speed is of distinct length. After the segmentation process, bi-spectrum of each slice was calculated to locate the peaks and its harmonic in two-dimensional spectral space. In the end, the calculated bi-spectra are provided as an

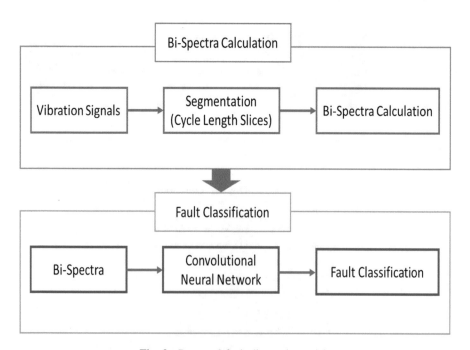

Fig. 2. Proposed fault diagnosis model.

input to the convolution neural network (CNN) so that it can apply layered vise filters and learn appropriate approximations to classify the data.

3.1 Signal Preprocessing

Let suppose $s(n)$ is real and a discrete stationary process with zero mean. It has third-order cumulant "$R_{ss}(t_1, t_2)$" which can be defined as:

$$R_{ss}(t_1, t_2) = E[s(n)s(n + t_1)s(n + t_2)] \tag{1}$$

So, the expression for the calculation of the bi-spectrum is given as:

$$R_{ss}(\omega_1, \omega_2) = \sum_{t_1=-\infty}^{+\infty} \sum_{t_2=-\infty}^{+\infty} R_{ss}(t_1, t_2) e^{-j(\omega_1 t_1 + \omega_2 t_2)} \tag{2}$$

Where, ω_1 and ω_2 are two frequency and t_1 and t_2 are two time variables used to calculate the bi-spectrum of a signal. Bi-spectrum of a signal basically decomposes the third order cumulant of the signal the way power of a signal is decomposed in the power spectrum.

3.2 Convolutional Neural Network

Convolution neural network CNN same as an ordinary neural network which is made up of neurons with learnable weights and biases. In CNN several convolution layers having non-linear activation functions are used. The typical activation function used in the convolution layers are rectified linear units (ReLU) or hyperbolic tangent function (tanh). In traditional artificial neural network (ANN) every neuron is connected to each neuron in the next layer, whereas in the CNN convolution operation is used to compute the output.

Architecture Overview

The architecture of the convolution neural network used in the proposed scheme is given in Fig. 3. It consists of six layers: Two convolution layers; two max-pooling layers; two fully connected layers. The convolution layers perform a convolution operation on the input data to calculate the output of each layer. The max-pooling layers are used to sub-sample the data coming from the previous layer. The input data provided to the input layer has 16382 dimensions, which passes through the network and hence in each layer filters are applied to learn the best approximation of the input data. In this way, the input data after every max-pooling operation of the network is sub-sampled by 2. In the given data set there are four classes, so the output of the final fully connected layer has four dimensions, each representing on class.

Layer (type)	Output Shape	Param #
conv1d_1 (Conv1D)	(None, 16382, 64)	256
max_pooling1d_1 (MaxPooling1	(None, 8191, 64)	0
conv1d_2 (Conv1D)	(None, 8189, 32)	6176
max_pooling1d_2 (MaxPooling1	(None, 4094, 32)	0
flatten_1 (Flatten)	(None, 131008)	0
dense_1 (Dense)	(None, 128)	16769152
dense_2 (Dense)	(None, 4)	516

Fig. 3. The Convolutional Neural network architecture used in this study.

4   Results

The endurance of the proposed model is tested on the vibration acceleration signals acquired from the rotary machine bearing under variable speed. The dataset contains vibration signals from four different shaft speeds, i.e., 1722, 1748, 1772, 1797 RPMs. The developed model is trained with 60% and tested with the remaining 40% of the data. In the pipeline, the first step was to perform the higher order spectral analysis of the signals through bi-spectrum calculation. The calculated bi-spectra for different health conditions of the bearings are presented through Figs. 4, 5, 6 and 7.

In Fig. 4 bi-spectrum of a signal representing normal health condition of the bearing is given. In bi-spectrum the third order cumulant of the signal is decomposed over two frequencies variables. In this way, it can reveal certain peaks through two-dimensional spectral analysis. From the figure, it can be observed that for the normal signal bi-spectrum certain frequency peaks are discernable. In this case, the dominant peaks occur in lower frequency bands.

In Fig. 5, bi-spectrum of an inner fault signal is given. It is evident that the graph of spectral quantities is simple and easy to interpret. The peeks of inner race fault and its harmonics are noticeable in the bi-spectra. The bi-spectrum identifies the transient variation among the faults in a better manner.

Figures 6 and 7 show the bi-spectra of an outer race and a rolling element fault signal, respectively. The frequency peaks and the harmonics are apparent in these cases as well. It is evident that the spectral smearing is not that much dominant. In these cases, the dominant peaks reside in the 0 to 0.4 Hz frequency band.

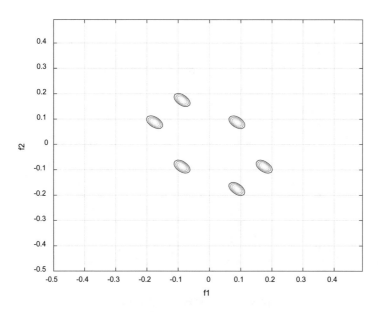

Fig. 4. Normal signal bi-spectrum.

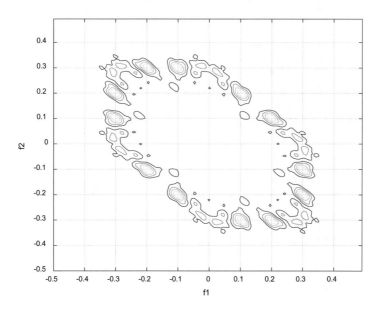

Fig. 5. Inner fault bi-spectrum.

Figure 8 presents the training and validation accuracy curves. It is apparent that both the curves are nicely converged. In both the cases, the initial accuracy is under 60% that increases rapidly with each epoch. The model achieves its best performance of more than 96% of training as well as validation accuracy near to 30 epochs.

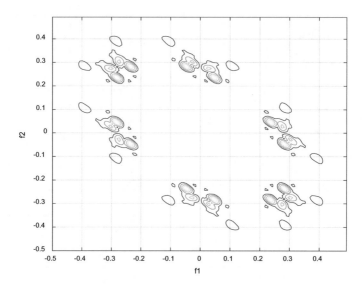

Fig. 6. Outer fault bi-spectrum.

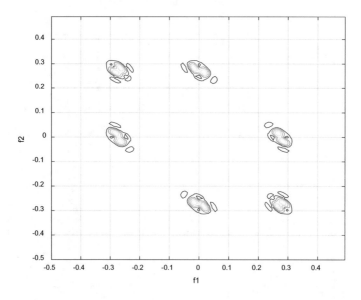

Fig. 7. Roller fault bi-spectrum

In Table 2, the comparison results of the proposed fault diagnosis scheme with other state of the art algorithms are given. The other schemes include feed forward artificial neural network (ANN) and vibration spectrum imaging (VSI) [15] used for

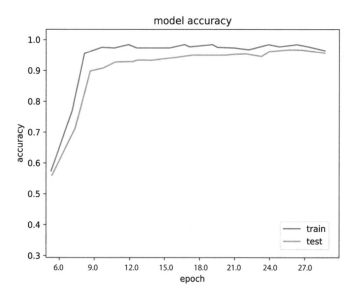

Fig. 8. Training and validation accuracy Curve.

Table 2. Comparison results of the proposed model and other machine learning algorithms for bearing fault diagnosis.

Algorithm	RPMs	Average accuracy
Feed Forward-ANN	1722, 1748, 1772, 1797	83
VSI	1722, 1748, 1772, 1797	85
Proposed	1722, 1748, 1772, 1797	98

fault diagnosis of the bearings. The preeminence of the proposed bearing fault diagnosis model is conspicuous over the rest of algorithms with 96% of average accuracy.

5 Conclusion

The current work presents a bearing fault diagnosis scheme based on convolutional neural network (CNN) and bi-spectrum analysis of the vibration signals. Bi-spectra of the vibration signals are calculated to extort the peaks in two-dimensional spectral space. The bi-spectra presented with some vibrant frequency peaks that made the fault identification and diagnosis easy for a given fault type with the help of the CNN. The bi-spectra are provided to the convolution neural network where layered vise filters are applied to learn the best possible approximation about the signals and classify the data into respective classes. Overall. the proposed scheme obtained an average classification accuracy above 90%. The results of the proposed method provide the evidence that it can be used for fault diagnosis of the bearing under variable speed conditions.

Acknowledgement. This work was supported by the Korea Institute of Energy Technology Evaluation and Planning (KETEP) and the Ministry of Trade, Industry & Energy (MOTIE) of the Republic of Korea (No. 20181510102160, No. 20162220100050, No. 20161120100350, and No. 20172510102130). It was also funded in part by The Leading Human Resource Training Program of Regional Neo industry through the National Research Foundation of Korea (NRF) funded by the Ministry of Science, ICT and future Planning (NRF-2016H1D5A1910564), and in part by the Basic Science Research Program through the National Research Foundation of Korea (NRF) funded by the Ministry of Education (2016R1D1A3B03931927).

References

1. Appana, D.K., Islam, M.R., Kim, J.-M.: Reliable fault diagnosis of bearings using distance and density similarity on an enhanced k-NN, pp. 193–203. Springer (2017)
2. Sohaib, M., Kim, J.-M.: A robust deep learning based fault diagnosis of rotary machine bearings. Adv. Sci. Lett. **23**, 12797–12801 (2017)
3. Wu, S., Chen, X., Zhao, Z., Liu, R.: Data-driven discriminative K-SVD for bearing fault diagnosis. In: 2017 Prognostics and System Health Management Conference (PHM-Harbin), pp. 1–6 (2017)
4. Liang, M., Su, D., Hu, D., Ge, M.: A novel faults diagnosis method for rolling element bearings based on ELCD and extreme learning machine. Shock. Vib. **2018**, 1–10 (2018)
5. Wei, Z., Wang, Y., He, S., Bao, J.: A novel intelligent method for bearing fault diagnosis based on affinity propagation clustering and adaptive feature selection. Knowl. Based Syst. **116**, 1–12 (2017)
6. Gao, Z., Cecati, C., Ding, S.X.: A survey of fault diagnosis and fault-tolerant techniques—Part I: fault diagnosis with model-based and signal-based approaches. IEEE Trans. Industr. Electron. **62**, 3757–3767 (2015)
7. Cecati, C.: A survey of fault diagnosis and fault-tolerant techniques—part II: fault diagnosis with knowledge-based and hybrid/active approaches. IEEE Trans. Ind. Electron. 3768–3774 (2015)
8. Wang, D., Peter, W.T., Tsui, K.L.: An enhanced Kurtogram method for fault diagnosis of rolling element bearings. Mech. Syst. Signal Process. **35**, 176–199 (2013)
9. Pineda-Sanchez, M., Perez-Cruz, J., Roger-Folch, J., Riera-Guasp, M., Sapena-Baño, A., Puche-Panadero, R.: Diagnosis of induction motor faults using a DSP and advanced demodulation techniques. In: 2013 9th IEEE International Symposium on Diagnostics for Electric Machines, Power Electronics and Drives (SDEMPED), pp. 69–76 (2013)
10. Appana, D.K., Ahmad, W., Kim, J.-M.: Speed invariant bearing fault characterization using convolutional neural networks, pp. 189–198. Springer (2017)
11. Kang, M., Kim, J., Kim, J.-M.: An FPGA-based multicore system for real-time bearing fault diagnosis using ultrasampling rate AE signals. IEEE Trans. Ind. Electron. **62**, 2319–2329 (2015)
12. Kang, M., Kim, J., Kim, J.-M.: High-performance and energy-efficient fault diagnosis using effective envelope analysis and denoising on a general-purpose graphics processing unit. IEEE Trans. Power Electron. **30**, 2763–2776 (2015)
13. Sohaib, M., Kim, J.-M.: Reliable Fault Diagnosis of Rotary Machine Bearings Using a Stacked Sparse Autoencoder-Based Deep Neural Network. Shock. Vib. **2018**, 11 (2018)
14. http://csegroups.case.edu/bearingdatacenter/home/
15. Amar, M., Gondal, I., Wilson, C.: Vibration spectrum imaging: A novel bearing fault classification approach. IEEE Trans. Ind. Electron. **62**, 494–502 (2015)

1D CNN-Based Transfer Learning Model for Bearing Fault Diagnosis Under Variable Working Conditions

Md Junayed Hasan, Muhammad Sohaib, and Jong-Myon Kim[✉]

Department of Electrical and Computer Engineering,
University of Ulsan, Ulsan, South Korea
junhasan@gmail.com, md.sohaibdurrani@gmail.com,
jmkim07@ulsan.ac.kr

Abstract. Classical machine learning approaches have made remarkable contributions to the field of data-driven techniques for bearing fault diagnosis. However, these algorithms mainly depend on distinct features, making the application of such techniques tedious in real-time scenarios. Under variable working conditions (i.e., various fault severities), the acquired signals contain variations in the signal amplitude values. Therefore, the extraction of reliable features from the signals under such conditions is important because it could discriminate the health conditions of the bearings. In this paper, a transfer learning approach based on a 1D convolutional neural network (CNN) and frequency domain analysis of the vibration signals is presented to solve the problem. Transfer learning enables the developed model to utilize information obtained under a given working condition to diagnose faults under other working conditions. The proposed approach has a classification accuracy of 99.67% when tested with the data acquired from the bearings with various fault severities. We also observe that a frequency spectrum enhances the performance of the transfer learning-based fault diagnosis model.

Keywords: Bearing fault identification · Convolutional neural network
Transfer learning · Vibration signals

1 Introduction

Bearing faults are some of the most prominent reasons for induction machine failure [1]. Incipient fault diagnosis of the bearings can avoid financial loss, reduce machine down time due to a failure, and improve plant production capacity [2, 3]. Bearing fault diagnosis can be divided into model-driven and data-driven techniques [4, 5]. Data driven techniques seem more promising for bearing fault diagnosis [1]. In such techniques, distinct features associated with the different health states of the bearing are extracted from signals without knowing the detailed architecture of the machinery. Later, the extracted features are provided to a sophisticated machine learning algorithm to classify the data into different classes.

However, the challenge in such a technique is to extract suitable features from the signals. Feature extraction is a tedious process, and the quality of the extracted features

© Springer Nature Switzerland AG 2019
S. Omar et al. (Eds.): CIIS 2018, AISC 888, pp. 13–23, 2019.
https://doi.org/10.1007/978-3-030-03302-6_2

affects the performance of the classical machine learning algorithms [6–8]. Moreover, the same types of data and feature space under a constant working condition are required for such methods [1]. Data collected from the bearing in the form of signals is a variable of working conditions, i.e., variation in a working condition effects the data values. In the case of bearings, fault severity, motor loads, and speeds are responsible for inconstant working conditions. The performance of the classical bearing fault diagnosis techniques deteriorates under variable working conditions due to the data variation. In addition, dealing with a massive amount of data makes the fault diagnosis task more difficult and effects the performance of a developed model. To address such problems, many deep learning techniques have been used in the field bearing fault diagnosis [2, 6, 9–11]. However, these methods suffer when dealing with big data in variable working conditions. To solve this issue, Zhang et al. [4] developed transfer learning using a convolutional neural network. The transfer learning process has been described in detail by using raw vibration signals for training and testing. However, many important points are left out, which are useful for fault classification. In another study by Wen et al. [1] a transfer learning model with an auto encoder was developed to solve the problem. However, the dimensionality reduction process through an auto encoder requires many additional efforts as well.

The main contribution of the current work is: (1) a transfer learning-based approach for fault diagnosis of the bearing with various fault severities, (2) a suggestion of a bearing fault diagnosis model that does not require well engineered features, and (3) introduction of a frequency spectrum calculation as a preprocessing step of the vibrations signals to explore the potential information and enhance the performance of the proposed model. To project a real-time scenario, we have created two different working conditions by using publicly available bearing data [12]. Both the working conditions have different fault severities and number of classes.

The rest of the paper is described as follows. Section 2 defines the proposed methods with the background, while Sect. 3 elaborates on the experimental setup and results of the experiment. Finally, Sect. 4 draws conclusions.

2   Proposed Method

2.1   Skimming Frame

In this study, vibration signals acquired from a bearing under two different working conditions are used for evaluation of the proposed bearing fault diagnosis model. In transfer learning, a network is first trained using the training data, and then the knowledge gained during the training is transferred to the target domain while testing the network. Hence, to create two working conditions, a huge amount of data is required. Thus, to create a significant number of samples, a skimming frame algorithm is used [4]. For each tiny step, one sample will be created.

If the total length of a vibration signal is "L", then the total number of the samples "n" will be:

$$n = \left(\frac{L - l}{s}\right) + 1. \tag{1}$$

In (1) "l" is the length of a single frame that needs to be selected based on the experimental requirement. The step size is considered as "s". The dimension of the vibration signal "L" is fixed. A complete process of the skimming frame algorithm is illustrated in Fig. 1.

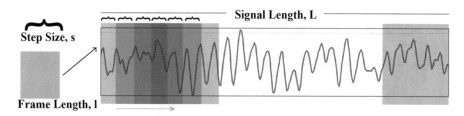

Fig. 1. Skimming frame process for generating samples.

2.2   Fast Fourier Transform

In this research, vibration signals are used in the experiments. It is easier to extract useful information from frequency domain by using Fast Fourier Transformation (FFT) [2]. If the number of samples is v, then FFT will consider $v * \log(v)$ operations. Figure 2 shows the FFT calculation process of the raw vibration signals. The FFT of each sample is calculated, and according to the Nyquist theorem, half of the data points in the frequency spectrum of each sample are considered for use further in the experiments.

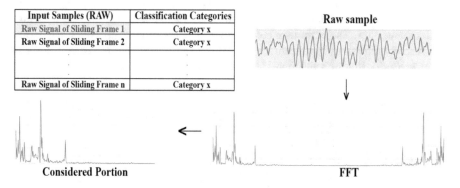

Fig. 2. FFT process for preprocessing the RAW signal.

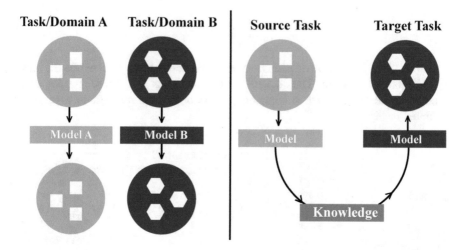

Fig. 3. (a) A traditional machine learning approach is presented which has two independent models for two different tasks; (b) a transfer learning approach is shown where the two tasks are denoted as source task and target task and knowledge is transferred between the tasks.

2.3   Transfer Learning

Developing a new network for a new task is a cumbersome and challenging process that adds an overhead to the fault diagnosis process. On the other hand, transfer learning makes the fault diagnosis process easy and robust [13]; it extracts valuable information from one working condition (task A) and applies that knowledge to the target working condition (task B) [4]. In general, transfer learning allows for learning from a vast number of past experiences, and then transferring that knowledge into different environments. In this research, first, the network learns distinct characteristics from massive amounts of source data in one working condition, and then passes that knowledge to a target task that has the same type of data but acquired under a different working condition. The main idea behind transfer learning is identifying a remarkable improvement regarding target scenarios from the knowledge that is gathered from the source scenario. Figure 3 illustrates the idea of transfer learning.

2.4   Overview of 1D CNN

A Convolution Neural Network (CNN) is one of the most effective supervised machine learning approaches [9]. In the current work, 1D CNN has been used in a transfer learning-based approach for machine fault diagnosis. CNN has a hierarchical architecture composed of convolutional, subsampling, and fully connected layers. 1D CNN is the same as conventional 2D CNN; the only difference is the dimension of the input data and filters used in the multiple layers. In 1D CNN, if the training data is $M = [m_1, m_2, \ldots \ldots m_j]$, then the number of training sample is j. Moreover, a target vector is $N = [n_1, n_2, \ldots \ldots n_j]$ which is associated with $M \cdot K$ layers constitute a CNN, then each layer in the network has F^K features, which are used in the convolution and subsampling processes [4]. If the sigmoid activation function is $\sigma(.)$, the weight matrix between input and hidden layer is w_1, the weight matrix between hidden and output layer is w_2, the bias vector of the hidden layer is b_1 and the output layer is b_2, then the feed forward process will be denoted as follows:

$$P = \sigma(w_1 M + b_1) \tag{2}$$

$$N = \sigma(w_2 P + b_2) \tag{3}$$

$$N = \sigma(w_2 P + b_2) \tag{4}$$

CNN is like an ordinary neural network that has adjustable weights and biases. In 1D CNN if the input signal length is L', then the input size is $(1 * L' * 1)$. As a result, the convolution layer will calculate the output of the neurons. In the proposed model, 32 1D filters have been considered in the convolution layer. After the convolution layer, a dropout layer is added to the network to avoid overfitting. The third layer in the network is max-pooling, which subsamples the data. The max-pooling layer is followed by a fully connected layer, which computes the class scores. Each neuron of the fully connected layer is connected to all the neurons in the previous layer. A stochastic gradient descent algorithm is used for the fine tuning of the network. A softmax classifier is added to the network for data classification. Figure 4 describes the overall architecture of the network.

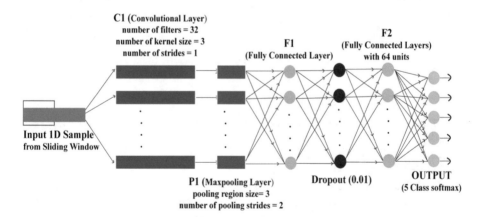

Fig. 4. Detail architecture of 1D CNN.

2.5 Details of the Proposed Approach

The proposed transfer learning-based bearing fault diagnosis consists of two phases. The first phase is the training phase of the network and the second is the testing phase of the model by transferring the domain knowledge. In this research, two working conditions are considered. First, the network is trained on the first working condition.

Then, the acquired knowledge is transferred to the second working condition. Figure 5 illustrates an overall workflow of the proposed model. The steps of the training phase are as follows:

1. Required source data is collected for the first working condition.
2. The skimming frame technique is applied to generate multiple samples.
3. To each sample, a label is assigned, as is necessary for supervised learning.
4. Fast Fourier Transform (FFT) calculation of the raw vibration samples is performed to obtain the frequency spectra of the signals.

Finally, the network (1D CNN) is trained using all samples from the training data, and the weights of the trained model are saved for later use during the testing phase.

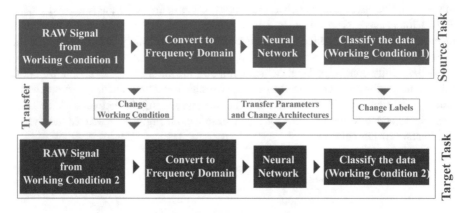

Fig. 5. Block diagram of the proposed model.

The Steps of the transferring phase are as follows:

1. As the working conditions are different, the data may contain atypical features and labels.
2. Labels are defined for each sample.
3. The FFT is calculated.
4. 2% of data is utilized to build the network architecture according to the data labels and dimensionality. The input size should be like the trained network, but the output will be different because of additional new labels.
5. The parameters of the main network will be unchanged.
6. Finally, the classification accuracy will be calculated.

The proposed model is evaluated by comparing the results with that of a fault diagnosis model where no preprocessing step is considered. Moreover, the results of the proposed model are also compared with the model where transfer learning is not used.

3 Experimental Setup

3.1 Dataset

To evaluate the proposed model, publicly available seeded fault bearing dataset by Case Western Reverse University was considered throughout the experiments. Drive end bearings were seeded with faults on the inner raceways, outer raceways, and rolling elements having a diameter of 0.007, 0.014, and 0.021 in. with the help of an electro-discharge machine as shown in Fig. 6.

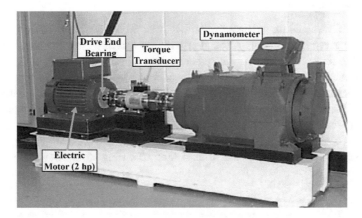

Fig. 6. Experimental set up by Case Western Reserve University [12].

Table 1. Details of the considered working conditions

Field name	Working condition 1	Working condition 2
Fault diameter	0.007 in.	0.021 in.
Motor load	0 horse power (hp)	0 horse power (hp)
Motor speed	1797 rounds/minute	1797 rounds/minute
Number of fault types	3	5
Number of labels	4	6
Name of the labels	Normal Inner fault Ball fault Outer fault at center	Normal Inner fault Ball fault Outer fault at center Outer fault at orthogonal Outer fault at oppositely
Utilized for	Source task	Target task

Thus, the dataset consists of an inner race fault, ball fault, and outer race fault signals under variable working conditions. Variable length vibration signals were recorded via an accelerometer attached to the housing of the motor close to the drive end bearing with a sampling data rate of 12,000 Hz. A dynamometer was used to generate different motor loads, ranging from 0 to 3 horsepower (hp). The placement and load zone on bearings create an impact on the vibration data of the outer race fault as it is stationary [12]. Details about the dataset and different working conditions are given in Table 1.

3.2   Result Analysis

The network is trained on the data from the first working condition. Table 1 shows that there are four types of signals associated with the health of the bearings for the first working condition. The four health states contain one normal and three faulty conditions. For each type, 3980 samples are considered. During the source training phase, the network gives 100% learning accuracy, while using 0.02% data for validation. The weights of the trained network are stored and transferred to the second working condition. For working condition 2, new labels are added to the dataset; thus, the output architecture of the network is modified. Therefore, from this working condition, 0.02% data is used to adjust the network architecture according to the new data. After the

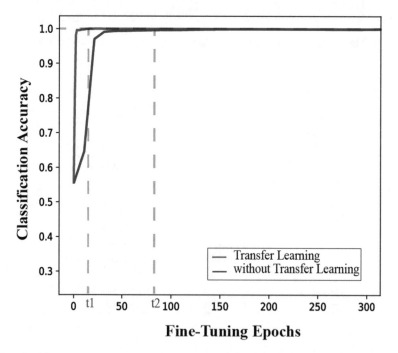

Fig. 7. Classification accuracy comparison of second working model (transfer learning-based model vs. without transfer learning).

network adjustment, the model is tested with the data of the second working condition. In addition, a separate CNN is developed and tested with the data of the second working condition without using the transfer learning technique, and the results are compared with that of the proposed model.

From Fig. 7, we see that both experiments can achieve around 100% accuracy. It is observed that by adopting a transfer learning technique in the fault diagnosis model, better performance can be achieved for the second working model at an earlier stage. After 16 fine tuning epochs (t1), the network can give almost 100% accuracy. On the other hand, without using transfer learning technique, after 80 fine-tuning epochs (t2), the network exhibits similar accuracy. Hence, the two approaches can achieve same performance in a different time span. In Table 2, the label-wise performance is given. For normal and ball fault condition, the improvement is 0%. The significance of using transfer leaning in the proposed model can be observed in case of the inner fault with an accuracy improvement of 4.8%. Overall, a 1.23% improvement in the average classification accuracy is achieved through a transfer learning-based approach for bearing fault diagnosis. In short, through transfer learning, better performance can be achieved at an earlier stage as compared to the conventional network.

Table 2. Comparison of classification accuracies of different approaches

Labels	Transfer learning	Without transfer learning	Improvements
Normal	100%	100%	0%
Inner fault	100%	95.2%	4.8%
Ball fault	100%	100%	0%
Outer fault at center	98.46%	98.13%	0.33%
Outer fault at orthogonal	100%	98.24%	1.76%
Outer fault at oppositely	99.54%	99.10%	0.44%
Overall	**99.67%**	**98.45%**	**1.23%**

To whether the preprocessing of the vibration signals has any impact on the performance of the proposed model, comparisons are made with the results of a 1D CNN that is trained on raw vibration signals. The training and testing approach of the network is identical to the proposed scheme: (1) train with the raw data of the first working condition, save, and then transfer the knowledge; (2) then test the model with the data from the second working condition. It is evident from Fig. 8 that with the same experimental set up, a preprocessing step can improve average classification accuracy of the model by at least 16%.

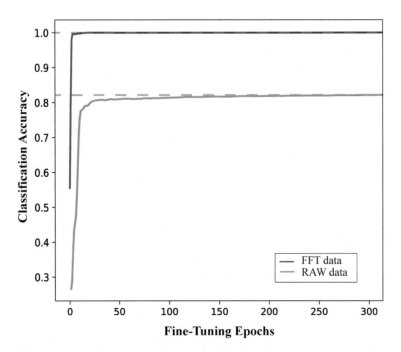

Fig. 8. Classification accuracy of the proposed model using preprocessed data (FFT data) vs. RAW data.

4 Conclusion

This paper proposed a novel fault diagnosis approach for the bearings of a rotary machine using a 1D convolutional neural network (CNN)-based transfer learning technique. Moreover, a preprocessing step was added into the fault diagnosis pipeline, which consists of a frequency spectrum calculation of the vibration signals. The addition of a preprocessing step in the proposed model enhanced the overall performance by extracting useful information. The results demonstrated that the proposed method achieves a 99.67% accuracy by using vibration signals acquired form the bearings under different working conditions. The frequency spectrum calculation of the vibration signals improved the average accuracy of the proposed model by 16%. In the future, the presented work can be extended to build an automated online fault detection system.

Acknowledgement. This work was supported by the Korea Institute of Energy Technology Evaluation and Planning (KETEP), and the Ministry of Trade, Industry, & Energy (MOTIE) of the Republic of Korea (Nos. 20181510102160, 20162220100050, 20161120100350, 20172510102130). It was also funded in part by The Leading Human Resource Training Program of Regional Neo industry through the National Research Foundation of Korea (NRF) funded by the Ministry of Science, ICT, and future Planning (NRF-2016H1D5A1910564), and in part by the Basic Science Research Program through the National Research Foundation of Korea (NRF) funded by the Ministry of Education (2016R1D1A3B03931927).

References

1. Wen, L., Gao, L., Li, X.: A new deep transfer learning based on sparse auto-encoder for fault diagnosis. IEEE Trans. Syst. Man Cybern. Syst. 1–9 (2017)
2. Ince, T., Kiranyaz, S., Eren, L., Askar, M., Gabbouj, M.: Real-time motor fault detection by 1-D convolutional neural networks. IEEE Trans. Ind. Electron. **63**, 7067–7075 (2016)
3. Eren, L., Devaney, M.J.: Bearing damage detection via wavelet packet decomposition of the stator current. IEEE Trans. Instrum. Meas. **53**, 431–436 (2004)
4. Zhang, R., Tao, H., Wu, L., Guan, Y.: Transfer learning with neural networks for bearing fault diagnosis in changing working conditions. IEEE Access. **5**, 14347–14357 (2017)
5. Gao, Z., Cecati, C., Ding, S.X.: A survey of fault diagnosis and fault-tolerant techniques Part I: fault diagnosis. IEEE Trans. Ind. Electron. **62**, 3768–3774 (2015)
6. Worden, K., Staszewski, W.J., Hensman, J.J.: Natural computing for mechanical systems research: a tutorial overview. Mech. Syst. Signal Process. **25**, 4–111 (2011)
7. Kang, M., Islam, M.R., Kim, J., Kim, J.M., Pecht, M.: A hybrid feature selection scheme for reducing diagnostic performance deterioration caused by outliers in data-driven diagnostics. IEEE Trans. Ind. Electron. **63**, 3299–3310 (2016)
8. Sohaib, M., Kim, C.-H., Kim, J.-M.: A hybrid feature model and deep-learning-based bearing fault diagnosis. Sensors **17**, 2876 (2017)
9. Malek, S., Melgani, F., Bazi, Y.: One-dimensional convolutional neural networks for spectroscopic signal regression. J. Chemom., 1–17 (2017)
10. Appana, D.K., Ahmad, W., Kim, J.-M.: Speed invariant bearing fault characterization using convolutional neural networks. In: Phon-Amnuaisuk, S., Ang, S.-P., Lee, S.-Y. (eds.) Multi-disciplinary Trends in Artificial Intelligence, pp. 189–198. Springer, Cham (2017)
11. Jia, F., Lei, Y., Lin, J., Zhou, X., Lu, N.: Deep neural networks: a promising tool for fault characteristic mining and intelligent diagnosis of rotating machinery with massive data. Mech. Syst. Signal Process. **72–73**, 303–315 (2016)
12. Case Western Reserve University: Bearing Data Center Website. http://csegroups.case.edu/bearingdatacenter/pages/download-data-file
13. Michalik, T., Polska, O.: How effective is Transfer Learning method for image classification, vol. 12, pp. 3–9 (2017)

Botnet Detection Using a Feed-Forward Backpropagation Artificial Neural Network

Abdulghani Ali Ahmed$^{(\boxtimes)}$

Systems Network & Security (SysNetS) Research Group,
Faculty of Computer Systems & Software Engineering,
Universiti Malaysia Pahang, 26300 Kuantan, Malaysia
abdulghani@ump.edu.my

Abstract. Botnet represent a critical threat to computer networks because their behavior allows hackers to take control of many computers simultaneously. Botnets take over the device of their victim and performs malicious activities on its system. Although many solutions have been developed to address the detection of Botnet in real time, these solutions are still prone to several problems that may critically affect the efficiency and capability of identifying and preventing Botnet attacks. The current work proposes a technique to detect Botnet attacks using a feed-forward backpropagation artificial neural network. The proposed technique aims to detect Botnet zero-day attack in real time. This technique applies a backpropagation algorithm to the CTU-13 dataset to train and evaluate the Botnet detection classifier. It is implemented and tested in various neural network designs with different hidden layers. Results demonstrate that the proposed technique is promising in terms of accuracy and efficiency of Botnet detection.

Keywords: Botnet · Feed-forward · Artificial Neural Network
Backpropagation

1 Introduction

In our fast-paced world driven by high technology, computers and other devices have become an essential part of our daily lives, and computerized systems have brought considerable convenience. However, with the rise of technology, security has become an issue for users. At present, the security of user information is threatened by viruses and other malicious attacks. The files affected by these viruses can severely affect applications in systems and other system files. The result is a complete breakdown of computer systems [1].

In recent times, Botnet detection has emerged as a popular research topic in the fields of cyber threats and cybercrimes. Botnet are considered one of the dangerous types of Botnet in network-based attacks. Botnet behavior allows hackers to take control of many computers simultaneously and to turn these computers into "zombies," as illustrated in Fig. 1. These "zombie computers" then operate as powerful Botnet that perform malicious activities, such as identity theft, distributed denial of service attacks, phishing, spamming, and domain name system spoofing.

© Springer Nature Switzerland AG 2019
S. Omar et al. (Eds.): CIIS 2018, AISC 888, pp. 24–35, 2019.
https://doi.org/10.1007/978-3-030-03302-6_3

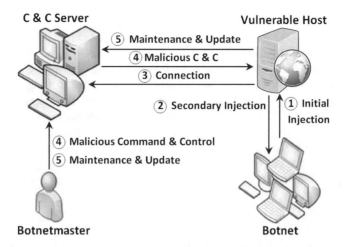

Fig. 1. Typical Botnet life cycle

This work reviews and analyzes several techniques used to detect Botnet attacks in computer systems. A large body of research [2–5] has explored the detection of Botnet activities in computer network systems. Many of these studies tend to apply recent machine learning algorithms, such as support vector machine (SVM) [6], decision tree [7], naïve Bayes (NB) [8], bees [9, 10], and random forest [11]. The feed-forward backpropagation algorithm [12] for training artificial neural networks (ANNs) to detect Botnet attacks is rarely studied and has yet to be evaluated comprehensively.

The present study introduces two revolutionary concepts. First, a feed-forward backpropagation ANN is examined to determine whether it is efficient and applicable to Botnet attack detection. Second, the efficiency and capability of the proposed technique in Botnet detection is evaluated by applying a backpropagation algorithm to the CTU-13 dataset [13] using multiple neural network (NN) designs with different hidden layers. The remaining parts of this paper are organized as follows. Section 2 discusses related works. Section 3 describes the feed-forward backpropagation ANN technique. Section 4 explains the implementation setting and presents the experiment results. Section 5 provides the conclusion and recommendations for future work.

2 Related Works

Numerous techniques and methods have been proposed to detect the existence of botnet attacks in computer network systems. A research [14] used an anomaly-based botnet detection method and aimed to detect botnet controllers by monitoring transport layer data. The technique is able to detect IRC botnet controllers running on any random port without requiring known signatures or captured binaries. It is entirely passive, and therefore, invisible to operators, scalable to large networks, and able to identify botnets that actually impact end users. This approach also has the capability to quantify the size of botnets and to identify and characterize their activities without actually joining botnets. The approach can detect bots using encrypted and obfuscated protocols.

Similarly, the Botsniffer [15] uses network-based anomaly detection to identify botnet command and control (C&C) channels in a local area network without any prior knowledge of botnet signatures. This technique is based on the observation that bots within the same botnet are likely to demonstrate spatiotemporal correlation and similar behavior, such as that of a protocol layer, because of preprogrammed activities related to C&C communication. Botsniffer captures the behavioral patterns of botnet traffic and utilizes statistical algorithms to detect botnets. It was primarily developed to detect centralized IRC types of botnet architecture. Therefore, it applies several correlation analysis algorithms to detect spatiotemporal correlations in network traffic with a low false positive rate.

BotDigger [16] is a system that utilizes fuzzy logic to define logical rules based on selected statistical facts and significant features that identify botnet actions. BotDigger deals with human reasoning and decision-making processes by applying fuzzy logic; it can then measure the influence of fuzzy membership sets. The efficiency of fuzzy logic in different applications depends on the suitable selection of the number, type, and parameter of the fuzzy membership of functions and rules.

Host-based botnet detection [17] was developed using a flow-based detection method by correlating multiple log files installed on host machines and segregating botmaster commands into different categories. Bots typically respond faster than humans do; hence, mining and correlating multiple log files can be easily realized. The technique can be efficiently performed for both IRC and non-IRC bots by correlating several host-based log files for certain C&C traffic detection. It can even detect C&C-type communications even if no payload is detected. Thus, the technique can be applied to non-IRC botnets.

Several studies have recently proposed to recognize and detect botnets using machine learning techniques. Decision tree [7] is a popular choice for differentiating between botnet and non-botnet traffic. It is particularly used for classification and prediction because it concentrates on classification rules shown as decision trees, which are deduced from a group of disordered and irregular instances. In a top-down recursive manner, this technique looks at qualities between the internal nodes of decision trees, judges the downward branches according to the different attributes of the nodes, and draws a conclusion from the leaf nodes in the decision trees. Thus, root-to-leaf nodes correspond to a conjunctive rule, and the entire tree corresponds to a group of disjunctive expression rules.

The advantage of the decision tree classification algorithm is that it produces rules that are easy to understand and deals with different data types without requiring a large computation. The decision tree can identify which attributes are considered highly significant. Simultaneously, it also suffers from certain deficiencies, including the difficulty of estimating continuation fields. This algorithm needs to perform extensive pre-treatment for chronological data. When the number of categories is excessive, errors may arise.

The NB [8] classifier is another machine learning-based algorithm that is considered an effective but simple classifier that is used in several applications to process natural language and retrieve information, among others. This algorithm is based on the

Bayesian theorem and is highly suitable when the dimensionality of inputs is high. This algorithm assumes that the effect of a variable value on a given class is independent of the value of another variable. For example, the NB inducer can compute the conditional probabilities of classes and choose the class with the uppermost posterior. The NB classifier can be trained effectively as a supervised machine learning algorithm depending on the precise nature of the probability model.

SVM [6] is a method for supervised pattern classification that has been successfully implemented in the pattern recognition field. SVM is an algorithm in machine learning for training classification and regression rules from desired datasets. It is known as the most preferable and suitable algorithm given its efficiency when dealing with high dimensionality feature spaces. This algorithm features a solid mathematical foundation, and the results it generates are simple but effective.

In summary, machine learning-based studies have made considerable progress in terms of botnet detection. In the current work, we use a feed-forward backpropagation ANN to train NNs for botnet attack detection. In particular, this study aims to evaluate the efficiency of the backpropagation algorithm in detecting botnet attacks compared with existing machine learning-based algorithms.

3   Feed-Forward Backpropagation for Botnet Detection

To detect Botnet attacks, we propose an intelligent technique that classifies network traffic as either malicious or non-malicious using feed-forward backpropagation ANN. In particular, the proposed technique is composed of two main components: the pre-processing component and the classifier component (Fig. 2).

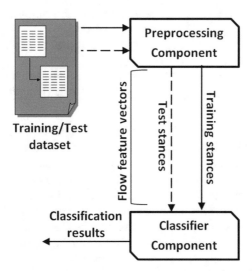

Fig. 2. Feed-forward backpropagation technique for Botnet detection

The preprocessing component analyzes network traffic on a flow level. Thus, a set of different features is extracted and selected for every traffic flow. The key aspect of flow-based traffic analysis is selecting a set of features of flow that efficiently helps identify targeted Botnet attacks. To this end, 15 statistical features are extracted for every traffic flow, as presented in Table 1.

Table 1. List of features extracted for every traffic flow

No.	Features	Description
1	Start time	Starting time of traffic flow
2	Duration	Total time taken to complete a particular flow
3	Protocol	The types of protocols used are TCP, ICMP, UDP, and SMTP
4	Source and destination IP addresses	From where the packet is sent and to where it is sent
5	Direction	A path of packet travels from source to destination
6	Source and destination ports	Identifying the data service or the location where the request should be sent
7	State	Flow state, e.g., SYN, RST, CON, ACK, and FIN
8	Type of Service (ToS)	Assigning a specific treatment or priority to each IP packet
9	Total packets	Number of packets in a specific flow or number of packets transmitted within a specific flow/time
10	Total bytes	Total number of bytes that the client sent per request
11	Time comparison	Comparison between flow start time and flow end time
12	Average byte rate	Average byte rate is calculated using basic features of total bytes and duration
13	Average packet rate	Average packet rate is calculated using basic features of total packets and duration
14	Ping byte	A packet with a size larger than 65435 bytes is considered malicious
15	Malicious port	Port number is also used to obtain information on remote systems that may be a target of malicious attacks

The common features are regarded to be insufficient to differentiate Botnet traffic from normal traffic [8]. Thus, to make the detection easy, we create new features such as average byte rate, average packet rate, ping bytes, time comparison, and malicious ports. These new features may help the system detect and determine Botnet traffic.

The classifier component is responsible for developing the model for Botnet/non-Botnet traffic based on the features extracted and selected by the preprocessing component. It then classifies traffic flow accordingly. The classification process operates in two main phases: training and test phases. In the training phase, the backpropagation learning algorithm is selected, and data are represented to develop the model. The backpropagation algorithm then generates a model that maps inputs onto targeted outputs. The backpropagation learning algorithm trains the generated model to predict the output for certain future inputs using feed-forward and backward processes [18], as illustrated in Fig. 3.

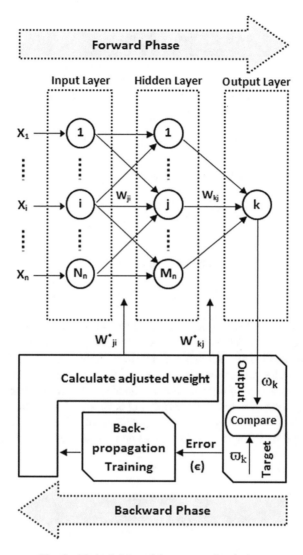

Fig. 3. Methodology of the proposed technique

Considering a feed-forward back propagation ANN-based network with x_n inputs (i) nodes, h hidden (j) nodes and o output (k) nodes, the back propagation training cycle passes through two main phases forward and backward phases. During forward propagation phase, a set of inputs vectors $x_1, ..., x_n$ are propagated by multiplying them with associated weights $w_1, ..., w_n$. Thus results of the output of prior nodes are multiplied with their respective weights and summed to calculate the net input to j^{th} node in the hidden layer as

$$net_j = \sum_{i=1}^{n} w_{ji} x_i \tag{1}$$

The obtained value from (1) specifies the output of ANN neurons and becomes the input value for the neurons in the next layer linked to it. Thus, the output (activation) of jth node in the hidden layer is given by:

$$o_{hj} = f(net_j) \tag{2}$$

Net input to k^{th} output node is calculated as

$$net_k = \sum_{j=1}^{L} w_{kj} o_{hj} \tag{3}$$

Net output ω_j to k^{th} output node is calculated as

$$\omega_j = f(net_k) \tag{4}$$

During the backward phase, the error signal is propagated through the network in the backward direction to adjust the weights and the bias values. The calculated weight changes are then applied to the free parameters of the network. The subsequent iteration starts, and the entire process is repeated using the next training model. The main purpose of this phase is to minimize the error signal in a statistical sense. The delta term for each output node ϵ_k is given by calculating error signal for each output node Δ_{ok} (the difference between the targeted value ϖ_k and actual values ω_k in the output layer) and multiplying it by the actual output of that node multiplied by (1-its actual output).

$$\Delta_{ok} = (\varpi_k - \omega_k)$$

$$\varepsilon_{ok} = \Delta_{ok} \omega_k (1 - \omega_k)$$

$$\varepsilon_{ok} = (\varpi_k - \omega_k) \omega_k (1 - \omega_k) \tag{5}$$

To calculate the error signal for each hidden node Δ_{hj}, summation of deltas of the output nodes connected to a particular hidden node is multiplied by the weight which connects that output and hidden nodes.

$$\Delta_{hj} = \sum_{k=1}^{w} \varepsilon_{ok} w_{kj}$$

The error signal for j^{th} hidden node is then multiplied by its output and by (1- its output) to get the delta term for j^{th} hidden node ϵ_{hj},

$$\varepsilon_{hj} = (o_{hj})(1 - o_{hj}) \sum_{k=1}^{W} \varepsilon_{ok} \, w_{kj} \qquad (6)$$

To get the weight error derivatives for each weight vector between hidden and output nodes γ_{jk}, delta of each output node is multiplied by the output (activation) of the hidden node connected to. The γ_{jk} is used to adapt the weight between the output and hidden layers as below.

$$\gamma_{jk} = \varepsilon_{ok}(o_{hj}) \qquad (7)$$

Likewise, the weight error derivatives for each weight between the input node and hidden node γ_{ij} are given by multiplying the delta of each hidden node with the activation of the input node it links to. The γ_{ij} is used to adapt the weights between the input and hidden layers.

$$\gamma_{ij} = \varepsilon_{hj}(x_i)$$

To control process of updating the weights during each backpropagation cycle, a learning rate parameter σ is needed to do the changes in these weights themselves. The weights on links between the hidden and output nodes at a time $(t + 1)$ are given using the weights at a time (t) and the γ_{jk} using the following equation.

$$w_{jk}(t+1) = w_{jk}(t) + \sigma(\gamma_{jk}) \qquad (8)$$

Likewise, the weights on links between the input and hidden units are given using the following equation.

$$w_{ij}(t+1) = w_{ij}(t) + \sigma(wed_{ij}) \qquad (9)$$

This way, every node in the ANN receives an error signal that shows its proportional contribution to the total of error between the targeted output and actual output. Thus, the update process on the weights linking the nodes in the different layers depends on the error signal these nodes obtained. By iterating the two processes in (8) several times for different input patterns and their targets, the mean square error between the actual output of the ANN and its desired output is reduced for all the set members of training inputs.

4   Implementation and Results

This section explains the implementation of the feed-forward backpropagation ANN technique used in this work to detect Botnet attacks. The implementation consists of several steps for dataset selection, feature extraction, data normalization, training, validation, and testing, as illustrated in Fig. 4.

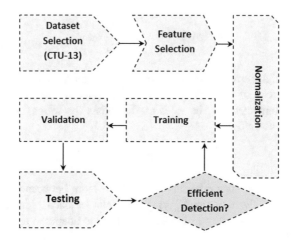

Fig. 4. Implementation of the proposed technique

In the first step, the CTU-13 dataset from the Botnet Capture Facility Project is used. The CTU-13 is a preprocessed dataset consists of 13 captures, also known as scenarios of different Botnet samples. These data are already labeled with normal traffic and Botnet traffic background. The next step is to identify the features to be selected as the input layer. The raw attributes of this dataset are shown in Table 2. Once the features have been selected, the data are normalized between the values of 0 and 1. Normalization is a critical step because it converts all numbers to a range between 0 and 1. This step ultimately results in a successful classification of data, in which 0 is considered normal traffic and 1 is Botnet traffic.

Table 2. Input layer

Input layer	Data attribute
X1	Total bytes
X2	Total packets
X3	Duration
X4	Source IP address
X5	Destination IP address
X6	Average bytes
X7	Average packets
X8	Source port
X9	Destination port

Once the data are normalized, the dataset undergoes three consecutive processes of training, validation, and testing. The processes of training, validation, and testing are implemented in MATLAB 2016 version 9. In this experiment, a total of 10000 randomly selected flows are distributed for the purposes of training, validation, and testing (Table 3).

Table 3. Flow distribution

Flows	Purpose
3000	Training
3500	Validation
3500	Testing

This distribution of data is applied to three different designs of NNs (Table 4). The numbers of input layers are similar for each design. The numbers of flows are also similarly distributed for each design.

Table 4. NN designs

ANN design	No. of hidden neurons
1	6
2	8
3	10

This section describes the performance and result of each NN design. Figure 5 shows that each hidden layer variation performs differently in terms of training, validation, and testing. The obtained results demonstrate that the number of hidden layers does contribute to the accuracy level. NN Design 3 with 10 hidden neurons exhibits the highest detection accuracy among all the designs. In general, after numerous training iterations, the mean square errors decrease, but are likely to increase in the validation

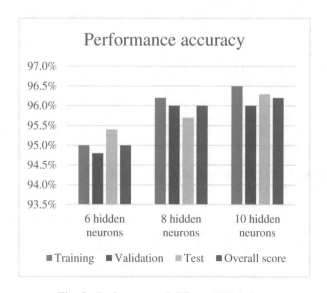

Fig. 5. Performance of different NN designs

dataset as the network starts overfitting the training data. Overfitting refers to a design that describes random errors or noise instead of the underlying relationship.

5 Conclusion

The feed-forward backpropagation ANN technique proposed in this work was found to be an effective and viable option that might be leveraged to detect Botnet attacks. The technique can improve the accuracy of NNs by manipulating the network hidden layer. Ensuring a reliable dataset also contributes significantly to the improvement of the performance of the feed-forward backpropagation algorithm. As demonstrated in this study, leveraging backpropagation for Botnet detection achieves a respectable accuracy of over 95%. This result indicates that feed-forward backpropagation ANN is promising. Researchers are recommended to conduct further studies on this algorithm to highlight its efficiency in detecting network threats such as Botnet attacks. As a future work, we are planning to apply Metaheuristic algorithm to find the best set of weights to be utilized in our proposed feed-forward back propagation ANN as a way to increase the accuracy level.

Acknowledgments. This work was supported in part by the Faculty of Computer System and Software Engineering, Universiti Malaysia Pahang under FRGS Grant No. RDU160106 and RDU Grant No. RDU160365.

References

1. Shah, S., Jani, H., Shetty, S., Bhowmick, K.: Virus detection using artificial neural networks. Int. J. Comput. Appl. **84**(5), 0975–8887 (2013)
2. Ahmed, A.A.: Investigation model for DDoS attack detection in real-time. Int. J. Soft. Eng. Comput. Syst. **1**(1), 93–105 (2015)
3. Ahmed, A.A., Jantan, A., Wan, T.-C.: Real-time detection of intrusive traffic in QoS network domains. IEEE Secur. Priv. **11**(6), 45–53 (2013)
4. Ahmed, A.A., Jantan, A., Wan, T.-C.: Filtration model for the detection of malicious traffic in large-scale networks. Comput. Commun. **82**, 59–70 (2016)
5. Ahmed, A.A., Jantan, A., Rasmi, M.: Service violation monitoring model for detecting and tracing bandwidth abuse. J. Netw. Syst. Manage. **21**(2), 218–237 (2013)
6. Narang, P., Ray, S., Hota, C., Venkatakrishnan, V.: Peershark: detecting peer-to-peer botnets by tracking conversations. In: Security and Privacy Workshops (SPW), IEEE, pp. 108–115 (2014)
7. Dai, Q., Zhang, C., Wu, H.: Research of decision tree classification algorithm in data mining. Int. J. Database Theory Appl. **9**(5), 1–8 (2016)
8. Kalaivani, P., Vijaya, M.S.: Mining based detection of botnet traffic in network flow. IRACST—Int. J. Comput. Sci. Inf. Technol. Secur. (IJCSITS) **6**(1), 2249–9555 (2016)
9. Jantan, A., Ahmed, A.A.: Honeybee protection system for detecting and preventing network attacks. J. Theor. Appl. Inf. Technol. **64**(1), 38–47 (2014)
10. Jantan, A., Ahmed, A.A.: Honey bee intelligent model for network zero day attack detection. Int. J. Digit. Content Technol. Appl. **8**(6), 45–52 (2014)

11. Singh, k, Guntuku, S.C., Thakur, A., Hota, C.: Big data analytics framework for peer-to-peer botnet detection using random forests. Inform. Sci. **278**, 488–497 (2014)
12. Svozil, D., Kvasnicka, V., Pospichal, J.: Introduction to multi-layer feed-forward neural networks. Chemometr. Intell. Lab. Syst. **39**(1), 43–62 (1997)
13. Garcia, S., Grill, M., Stiborek, J., Zunino, A.: An empirical comparison of botnet detection methods. Comput. Secur. **45**, 100–123 (2014)
14. Karasaridis, A., Rexroad, B., Hoeflin, D.A.: Wide-scale botnet detection and characterization. HotBots **7**, 7 (2007)
15. Gu, G., Zhang, J., Lee, W.: BotSniffer: Detecting Botnet Command and Control Channels in Network Traffic. In: NDSS, vol. 8, pp. 1–18 (2008)
16. Al-Duwairi, B., Al-Ebbini, L.: BotDigger: A fuzzy inference system for botnet detection. In: 2010 Fifth International Conference on Internet Monitoring and Protection (ICIMP), pp. 16–21. IEEE (2010)
17. Masud, M.M., Al-Khateeb, T., Khan, L., Thuraisingham, B., Hamlen, K.W.: Flow-based identification of botnet traffic by mining multiple log files. In: First International Conference on Distributed Framework and Applications, DFmA, pp. 200–206. IEEE (2008)
18. Rumelhart, D.E., Durbin, R., Golden, R., Chauvin, Y.: Backpropagation: the basic theory. In: Backpropagation: Theory, Architectures and Applications, pp. 1–34 (1995)

Modified ANFIS with Less Model Complexity for Classification Problems

Noreen Talpur, Mohd Najib Mohd Salleh$^{(\boxtimes)}$, Kashif Hussain,
and Haseeb Ali

Faculty of Computer Science and Information Technology,
Universiti Tun Hussein Onn Malaysia, 86400 Batu Pahat, Johor, Malaysia
najib@uthm.edu.my
http://fsktm.uthm.edu.my/

Abstract. A machine learning technique develops the best-fit model by adjusting weights based on learning from data. Similarly, adaptive neuro-fuzzy inference system (ANFIS) is also one of the commonly used machine learning techniques which employs training algorithm to adjust its parameters to approximate the problem under consideration. Even though, ANFIS is used in wide variety of applications including rule-based control systems, classification, and pattern matching, but ANFIS has drawback of computational complexity as it carries the problem of curse of dimensionality. This limits ANFIS to be used only with the applications having less number of inputs. Additionally, the gradient-based learning algorithm suffers from the problem of trapping in local minima. To address these drawbacks, this study reduces ANFIS architecture from five to four layers, in order to reduce model complexity. Moreover, to avoid the local minima problem in typical hybrid learning algorithm, the popular swarm-based metaheuristic algorithm Artificial Bee Colony (ABC) is used to solve the benchmark classification problems with varying input-size. The overall comparison of results and experiments show that the modified ANFIS model performed equally better as compared to standard ANFIS, but with significantly reduced trainable parameters and training computation cost. The modified ANFIS reduced the model complexity up to 93% on classification problems with large input-size.

Keywords: ANFIS · Neuro-fuzzy · Hybrid algorithm
Metaheuristic algorithms · Artificial bee colony

1 Introduction

Machine learning techniques such as artificial neural networks, fuzzy systems, genetic programming and neuro-fuzzy systems have been widely used approaches for solving nonlinear real-world problems. These techniques use computational intelligence to analyze and learn from vast amounts of digital data through iterative learning [1]. The applications include education, finance and economics, medical science, electronics, traffic controlling, image processing, manufacturing, forecasting and predictions, and social sciences [2].

© Springer Nature Switzerland AG 2019
S. Omar et al. (Eds.): CIIS 2018, AISC 888, pp. 36–47, 2019.
https://doi.org/10.1007/978-3-030-03302-6_4

Fuzzy logic is capable of dealing with uncertainty in data with the help of IF-THEN rules [3]; however, it is unable to learn. On the other hand, artificial neural network is able to learn from data but it cannot explain the decision, hence called "black-box" [4]. Subsequently, a hybrid of these two techniques offers a robust machine learning technique that is able to overcome the limitations of a single system [5]. One example of such hybrid is adaptive neuro-fuzzy inference system (ANFIS) [6] which incorporates learning mechanism of neural networks and reasoning approach of fuzzy logic for solving non-linear problems where it is difficult to build rules manually [7]. ANFIS uses gradient-based learning approach to training the entire model and minimizing error between actual and desired output.

Although, ANFIS has been widely used approach and merges the advantages of neural network and fuzzy system, the system faces serious computational complexity in terms of neurons (activation functions) to be computed, which hinders implementation on problems with large number of inputs. Furthermore, the standard hybrid learning algorithm used by ANFIS is inspired by the back-propagation algorithm of neural network, which has high possibility to falling in local minima [8]. To improve the learning of ANFIS, many derivative free training methods have been proposed by researchers in literature; such as, metaheuristics algorithms. Additionally, rule-base minimization techniques have been applied to reduce the computation complexity [9]. Yet, ANFIS model complexity remains a problem.

The aim of this study is to propose an ANFIS model with less computational complexity by modifying its standard architecture and instead of using typical hybrid leaning algorithm, this study uses one of the commonly used metaheuristics algorithms called artificial bee colony (ABC). The remaining paper is organized as follows. Section 2 analyzes and discusses the earlier work published by previous researchers regarding ANFIS and its training approaches. Section 3 explains the standard architecture of ANFIS model with the learning algorithm. The proposed approach is presented in research methodology part in Sect. 4 with the proposed ANFIS model. The basic setting of parameters and information about datasets is explained in Sect. 5. The results and discussion is presented in Sect. 6. Lastly, a summary of this study has been presented Sect. 7.

2   Related Work

The ANFIS algorithm is a hybrid of neural network and fuzzy logic, which has attracted researchers from various scientific and engineering areas. Because of robust results generated by ANFIS, it has been used in applications including rule-base control systems, classification and pattern matching [2].

Despite the fact that, ANFIS has produced better results as compare to other machine learning techniques like neural networks and fuzzy logic, yet the model faces some serious computational complexities. ANFIS uses grid partitioning method to generate rules with all possible combinations [10]. Therefore, the number of rules and parameters surges when the number of inputs in underlying system increases. This issue has been addressed and solved in literature by employing various rule-base minimization in ANFIS model. Those techniques include clustering approaches like optimized hyperplane clustering synthesis (OHCS), hierarchical hyperplane clustering synthesis (HHCS), subtractive clustering (SC), particle swarm optimization (PSO) optimized ANFIS and Karnap Map (K-Map) through selecting the rules with higher firing strength [9, 11–13].

ANFIS model becomes significantly complex when the number of rules rise remarkably due to the problem of large input-size. In fact, various real-world problems contain tens of inputs. Generally, the ANFIS model is suitable to problems with small input-size – up to 4 to 5 inputs [14]. Besides the complications of extensive rules and large inputs, the training of ANFIS parameters is also considered as serious problem. As the standard two-pass learning algorithm in ANFIS is inspired from back-propagation algorithm in neural networks, it is prone to finding suboptimal solutions. Alternatively, a number of studies have used metaheuristic algorithms to optimize ANFIS parameters. According to two comprehensive surveys conducted by Salleh and Hussain [15] and Karaboga and Kaya [1], some of the commonly used metaheuristic algorithms employed on ANFIS parameters training are genetic algorithms (GA), particle swarm optimization (PSO), and artificial bee colony (ABC). These techniques are either employed on premise part or consequent part of both. A detailed work on metaheuristic research is performed by Hussain et al. [16] found ABC as one of the top metaheuristic algorithms, which has produced promising results in variety of optimization problems.

According to literature, Rini and Basser [17, 18] used PSO to modify the antecedent and consequent parameters of ANFIS models. While, Bagheri [19] employed quantum-behaved particle swarm optimization (QPSO) for tuning premise parameters of ANFIS with the replacement of gradient decent algorithm for financial forecasting which was later improved by Liu [20] for training of premise plus consequent parameters of ANFIS via an improved version of quantum-behaved particle swarm optimization algorithm (QPSO). Nhu et al. [21] employed firefly algorithm (FA) with ANFIS as a time series forecasting system to predict stock market based on Hanoi Stock Exchange (HNX) in Vietnam. Similarly, [22, 23] employed ABC with ANFIS model to tune all the parameters of ANFIS. Whereas, Meysam et al. [24] proposed a new swarm-based metaheuristic optimization algorithm, known as cat swarm optimization (CSO), and employed on training the consequent parameters in ANFIS architecture. The CSO was used in hybrid of gradient descent method to tune membership function parameters.

To deal with inefficient learning, metaheuristic algorithms have been used. However, to address model complexity, there exists limited research. da Costa Martins and Arajo [25] modified ANFIS structure by independently leaving the inputs of the first and fifth layers. The proposed modification offered several advantages over standard ANFIS including application on problems with large input-size. Moreover, the model produced satisfactory model approximation results, however the ANFIS model still used five-layer architecture. Peymanfar et al. [26] also modified ANFIS structure for efficient learning. The research employed modified ANFIS on programing CMOS fuzzy controllers. In this work, the researchers removed third layer of standard ANFIS architecture, as this does not affect the operational rules of fuzzy logic. Additionally, the work has also proposed new learning algorithm that helped produce minimum error.

Based on gaps identified in literature, this study modifies ANFIS architecture to reduce its complexity, plus one of the popular metaheuristics approach artificial bee colony (ABC) is used to train parameters of the proposed ANFIS model.

3 Adaptive Neuro-Fuzzy Inference System (ANFIS)

ANFIS is a Sugeno-fuzzy type multilayer feedforward network which was introduced by Jang in 1993 [6]. The model has advantage of being more transparent to the user because of its ability to reveal rules that produce output. ANFIS has the capability of adaptability and handling problems with nonlinearity [27]. As illustrated in Fig. 1, the architecture of ANFIS comprises of five layers and two types of nodes (adaptive and fixed nodes). The nodes with square shapes are called adaptive nodes, hence, the parameters of those nodes are trained by hybrid algorithm. Nodes with circle shape are fixed nodes. ANFIS uses fuzzy *IF-THEN* rules as following:

$$\text{IF } x \text{ IS } \mu_A \text{ AND } y \text{ IS } \mu_B \text{ THEN } f = xp + yq + r$$

where μ_A and μ_B are fuzzy membership functions. The parameters p, q, and r are linear parameters that need to be identified during the process of training. The IF part of fuzzy rule maintains membership function parameters, whereas consequent parameters are in THEN part. In ANFIS architecture, the parameters in premise part (layer 1) and consequent part (layer 4) are trainable parameters. The nodes (activation functions) in rest of the layers are fixed and do not contain parameters. The ANFIS architecture with five layers is explained as:

Layer 1 (Fuzzification): The node i in this layer is adaptive membership function of any shape (Trapezoidal, Triangular, Gaussian, or generalized Bell shaped (1).

$$O_1, i = \mu_{Ai}(x), \mu_{Bi}(y), i = 1, 2 \tag{1}$$

where x and y are the two inputs with single output $O_{1,i}$ of the ith node.

Layer 2 (Product): This layer represents the product ($\prod$) to calculate firing strength of a rule. This layer accepts input values from first layer and turns into firing strength of ith rule as in (2).

$$O_2, i = w_i = \mu_{Ai}(x).\mu_{Bi}(y), i = 1, 2 \tag{2}$$

Layer 3 (Normalization): Each node in this layer normalizes firing strength of a rule by calculating the ratio of the ith rule's firing strength to the sum of all rules firing strength. The $\bar{w}_i$ is referred to as normalized firing strength of a rule (3).

$$O_3, i = \bar{w}_i = \frac{w_i}{w_1 + w_2}, i = 1, 2 \tag{3}$$

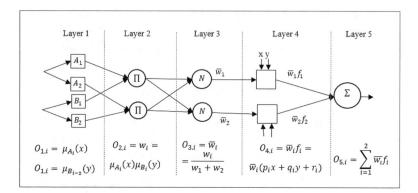

Fig. 1. ANFIS structure [6]

Layer 4 (Defuzzification): Parameters in this layer are linear parameters and identified during the training process (4).

$$O_4, i = \bar{w}_i f_i = \bar{w}_i(p_i x + q_i y + r_i), i = 1, 2 \qquad (4)$$

where $\bar{w}_i$ is rule's normalized firing strength and $p_i x + q_i y + r_i$ is a first order polynomial. O_4, i represents output of ith node in layer 4.

Layer 5 (Overall output): This layer contains single node which is labeled as Σ in Fig. 1. This layer performs summation of outputs in previous layer to generate single output (5).

$$O_5 = \sum_{i=1}^{2} \bar{w}_i f_i \qquad (5)$$

ANFIS learns by adjusting membership function parameters and consequent parameters (p, q, r) which are trainable parameters to minimize error between actual and desired output. During forward pass, ANFIS calculates nodes outputs until fourth layer by utilizing least squared estimation (LSE) to update consequent parameters before calculating the final output. In backward pass, error is propagated backward until first layer where ANFIS employs gradient descent (GD) to adapt membership function parameters [28].

4 Methodology

According to literature review, ANFIS is preferably suitable for problems with small number of inputs. Even though grid partitioning helps ANFIS produce better accuracy by generating maximum number of rules, but it also contributes to increase in computational cost – as consequent part of rules contains more number of trainable parameters. Therefore, the fourth layer maintains the most of model complexity. Moreover, various metaheuristic algorithms have been developed by researchers to address the issue of gradient-based learning. However, there is found limited research which solved the issue of model complexity by reducing or modifying the ANFIS architecture. Therefore, the core objective of this paper is to solve the problem of model complexity by reducing ANFIS architecture from five layers to four so that the ANFIS to training process has less parameters to process. As far as ANFIS training is concerned, this study employs one of the popular metaheuristic algorithm ABC instead of using hybrid learning algorithm. A brief introduction of ABC and its implementation on ANFIS parameter training is presented in the following subsection.

4.1 The Proposed ANFIS Model

Since, grid partitioning produces maximum number of rules serving ANFIS with better accuracy, but it also increases the computational cost as most of the trainable parameters are maintained in consequent part of the rules in fourth layer. Therefore, elimination of the fourth layer may reduce the training cost. Additionally, instead of tuning three parameters p, q, r in the consequent part of a rule, the proposed ANFIS architecture tunes single parameter r per rule, hence reducing the burden of trainable parameters. Thus, (4) and (5) are replaced with (6) and (7) respectively:

$$O_3, i = \bar{w}_i r_i = \frac{w_i}{w_1 + w_2} \times r_i, i = 1, 2 \tag{6}$$

$$O_4, i = \sum_{i=1}^{2} \bar{w}_i r_i, i = 1, 2 \tag{7}$$

where O_3, i is ith rule's output in third layer which is simply normalized weight times parameter r. The learning algorithm plays an important role in ANFIS accuracy. Therefore, as observed form the literature that several swarmbased metaheuristic algorithms have been used to train ANFIS models. Thus, this study reduces computational complexity of ANFIS architecture (Fig. 2), plus employs ABC on training ANFIS parameters.

ABC is one of the efficient and commonly used swarm-based metaheuristic algorithms, introduced by Karaboga in 2005 [29]. The swarm of population in ABC is divided into three groups: employed bees, onlooker bees, and scout bees, in order to divide the workforce effectively. Employed bees visit flower patches to determine nectar amount and share with onlooker bees waiting in bee hive. Onlooker bees choose flowers to visit based on information shared by employed bees. Scout bees, on the other hand, replace underperforming bees and visit unexplored locations in search of better nectar amount. Briefly, employed bees and onlooker bees are responsible for exploiting already identified promising locations, whereas scout bees are to explore the environment.

Since ABC is a population-based metaheuristic algorithm, each population individual maintains a trainable parameters of modified ANFIS architecture. These parameters include membership function parameters and one single coefficient in third layer (Fig. 3). In every iteration of ABC, each population individual is evaluated on objective function which contains the modified ANFIS model. The parameters are

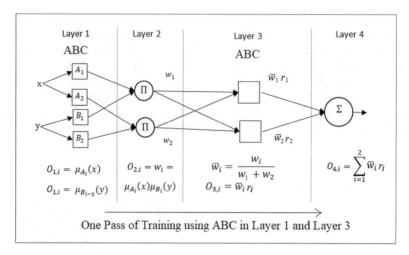

Fig. 2. Proposed ANFIS structure

applied on ANFIS model and error measure is calculated. Towards the end of iterations, ABC converges to the best error measure found so far. It is important to mention that this approach is flexible, as many other population-based metaheuristic algorithms can be employed in this methodology.

Population Individuals in ABC		
Pop. Ind.$_1$	**...**	**Pop. Ind.$_n$**
Parameters of Membership Function(s) and one Linear Coefficient	...	Parameters of Membership Function(s) and one Linear Coefficient

Fig. 3. ABC representation in ANFIS parameter training

5   Experiments

To achieve the objectives of this paper, experiments have been conducted for standard ANFIS architecture and the proposed modified ANFIS; both trained by the ABC algorithm. Performance of both models was compared in terms of training mean squared error (MSE) (8), number of trainable parameters and number of epochs by setting the error tolerance for overall output less than or equal to 0.005. According to the Table 1, eight benchmark classification problems were collected from UCI[1] and

---

[1] https://archive.ics.uci.edu/ml/datasets.html.

Table 1. Classification datasets

Dataset name	Number of attributes	Total number of instances
IRIS (D_1)	4	150
Teaching assistant evaluation (D_2)	5	151
Car evaluation (D_3)	6	1728
Seeds (D_4)	7	210
Breast cancer (D_5)	9	286
Glass identification (D_6)	10	214
Red wine quality (D_7)	11	1599
Statlog (D_8)	13	270

KEEL[2] repositories with inputs ranging from 4 to 13. These datasets were partitioned further into training and testing sets with the ratio of 70:30.

$$MSE = \frac{\sum_{i=1}^{m} \left(Output_{ANFIS} - Output_{Target} \right)^2}{m} \tag{8}$$

where m is the size of training dataset, $Output_{ANFIS}$ is ANFIS generated output and $Output_{Target}$ is the target output of mth training pair.

Since MATLAB is popular for its simplicity in programming and debugging mathematical operations. Therefore, all simulations and experiments were programmed using MATLAB 8.1.0.604 (R2013a) 64-bit version with the work environment Intel(R) Core i7 CPU 2.00 GHz, 2.50 GHz, 8 GB RAM and 64-bit Operating System.

As ANFIS model practices fuzzy logic, thus type and number of membership function per input plays a significant role in the performance of the model. According to [28], Gaussian type and two membership functions per input are suitable when employed in ANFIS with grid partitioning method. Hence, the experiments involved in this paper used Gaussian type and two membership functions with both standard and proposed ANFIS model. In order to measure computational complexity of both models, total number of trainable parameters were compared between standard ANFIS and modified ANFIS model by using (9):

$$F(n, m, p) = n \times m \times p + m^n \times (n + 1) \tag{9}$$

where n is the number of inputs and m represents the number of membership functions, whereas parameters need to be trained in membership function are labeled as p. Therefore, total number of premise trainable parameters are $n \times m \times p$. Similarly, the trainable parameters in consequent part per rule are $n + 1$ and m^n expresses total number of rules in system. Hence, the total number of consequent trainable parameters are $m^n \times (n + 1)$.

---

[2] http://sci2s.ugr.es/keel/datasets.php.

The ABC algorithm was used with both the ANFIS models to train the parameters using only one pass. The control parameters of ABC were set with the population size 50, $Limit = N \times D$, D = number of trainable parameters. Total number of iterations were 200.

6   Results and Discussion

In this section, the analysis was carried out carefully based on the performance of standard ANFIS and the modified ANFIS model trained by ABC.

According to Table 2, the training error of modified ANFIS is smaller than standard ANFIS on most of the classification datasets. The modified ANFIS achieved better accuracy on Iris (D_1), Teaching Assistant Evaluation (D_2), Seeds (D_4), Breast Cancer (D_5), and Red Wine Quality (D_7) problems. On the other hand, the standard ANFIS achieved slightly smaller MSE on Car Evaluation (D_3), Glass Identification (D_6), and Statlog (D_8). The training error measures of the modified ANFIS are encouraging, as it reveals elimination of fourth layer did not affect model performance. Moreover, instead of gradient-based learning, ABC trained the ANFIS parameters efficiently. Apart from training error, the model complexity of standard and the modified ANFIS can be evaluated by comparing the number of trainable parameters in Table 2.

Table 2. Simulation results of standard ANFIS and modified ANFIS

	Standard ANFIS			Modified ANFIS		
	MSE	Trainable parameters	Training cost (Seconds)	MSE	Trainable parameters	Training cost (Seconds)
D_1	0.0039	96	2.8161E+3	0.0031	32	1.5159E+3
D_2	0.1863	212	4.5884E+3	0.0400	52	3.5099E+3
D_3	0.0309	472	9.9088E+4	0.0337	88	8.2156E+4
D_4	2.50E−4	1,052	3.1746E+5	1.17E−4	156	1.3048E+5
D_5	0.0113	5,156	4.3050E+5	0.0102	546	1.7198E+5
D_6	0.0115	11304	4.9360E+5	0.0176	1064	1.3045E+5
D_7	0.0802	24620	5.1682E+5	0.0637	2092	2.0132E+5
D_8	0.1252	114740	1.0981E+6	0.1286	8244	8.3619E+5

The modified ANFIS was trained with significantly smaller number of trainable parameters as compared to standard ANFIS on all the classification problems. This infers that the modified ANFIS achieved reduction in model complexity by 67% on D_1, 76% on D_2, 81% on D_3, 85% on D_4, 89% on D_5, 91% on D_6, 92% on D_7, and 93% on D_8. It is clear from results that the more number of inputs, the more are the trainable parameters in standard ANFIS mode. On the other hand, it is reverse in case of modified ANFIS which significantly reduced the trainable parameters as the input-size increased. Therefore, the modified ANFIS has reduced the model complexity considerably. It is clear from the percentage that the proposed modified ANFIS can significantly reduce model complexity specially on classification problems with large input-size.

From the results of computation time or parameter training cost presented in Table 2, it is obvious that the proposed elimination of fourth layer and reduction in the number of parameters reduced time complexity in modified ANFIS. On all the classification problems, the standard ANFIS maintained higher computational cost as compared to modified ANFIS.

7   Conclusion

This study is an attempt to solve the problems of model complexity in the standard ANFIS algorithm. The complexity of ANFIS, in terms of trainable parameters, surges increasingly as the number of inputs enlarges. The proposed modified ANFIS architecture eliminated fourth layers of the standard ANFIS structure, as most of the trainable parameters reside in this layer. Instead of multiple linear coefficient parameters in the consequent part of fuzzy rules, the modified ANFIS used on single trainable parameter. The ABC algorithm was employed on training the membership function parameters as well as single parameter in the third layer. The ABC algorithm trained ANFIS models and produced better error measures on most of the classification problems. Our experiments showed that the proposed modification in ANFIS structure significantly reduced the number of trainable parameters, hence reduced model complexity and training cost in modified ANFIS. This encourages the applications of modified ANFIS on the problems with large input-size. Furthermore, the proposed methodology in this research can be handily employed on using other metaheuristic algorithms in place of ABC algorithm.

Acknowledgments. The authors would like to thank Universiti Tun Hussein Onn Malaysia (UTHM) for supporting this research under Postgraduate Incentive Research Grant, Vote No. U728 and Vote No. U560.

References

1. Karaboga, D., Kaya, E.: Adaptive network based fuzzy inference system (ANFIS) training approaches: a comprehensive survey. Artif. Intell. Rev., pp. 1–31 (2018)
2. Kar, S., Das, S., Ghosh, P.K.: Applications of neuro fuzzy systems: a brief review and future outline. Appl. Soft Comput. **15**, 243–259 (2014)
3. Walia, N., Singh, H., Sharma, A.: ANFIS: Adaptive neuro-fuzzy inference system-a survey. Int. J. Comput. Appl. **123**(13), 32–38 (2015)
4. Maind, S.B., Wankar, P., et al.: Research paper on basic of artificial neural network. Int. J. Recent Innov. Trends Comput. Commun. **2**(1), 96–100 (2014)
5. Akbari, S., Mahmood, S.M., Tan, I.M., Hematpour, H.: Comparison of neuro-fuzzy network and response surface methodology pertaining to the viscosity of polymer solutions. J. Pet. Explor. Prod. Technol., pp. 1–14 (2017)
6. Jang, J.-S.R.: ANFIS: adaptive-network-based fuzzy inference system. IEEE Trans. Syst. Man Cybern. **23**(3), 665–685 (1993)

7. Aghbashlo, M., Hosseinpour, S., Tabatabaei, M., Younesi, H., Najafpour, G.: On the exergetic optimization of continuous photobiological hydrogen production using hybrid ANFIS–NSGA-ii (adaptive neuro-fuzzy inference system–non-dominated sorting genetic algorithm-ii). Energy 96, 507–520 (2016)
8. Behmanesh, M., Mohammadi, M., Naeini, V.S.: Chaotic time series prediction using improved ANFIS with imperialist competitive learning algorithm. Int. J. Soft Comput. Eng. 4(4), 25–33 (2014)
9. Hussain, K., Salleh, M., Najib, M.: Analysis of techniques for anfis rule-base minimization and accuracy maximization. ARPN J. Eng. Appl. Sci. 10(20), 9739–9746 (2015)
10. Akrami, S.A., El-Shafie, A., Jaafar, O.: Improving rainfall forecasting efficiency using modified adaptive neuro-fuzzy inference system (manfis). Water Resour. Manag. 27(9), 3507–3523 (2013)
11. Rini, D.P., Shamsuddin, S.M., Yuhaniz, S.S.: Particle swarm optimization for anfis interpretability and accuracy. Soft. Comput. 20(1), 251–262 (2016)
12. Soh, A.C., Kean, K.Y.: Reduction of ANFIS-rules based system through k-map minimization for traffic signal controller. In: 2012 12th International Conference on Control, Automation and Systems (ICCAS), pp. 1290–1295. IEEE (2012)
13. Fattahi, H.: Indirect estimation of deformation modulus of an in situ rock mass: an ANFIS model based on grid partitioning, fuzzy c-means clustering and subtractive clustering. Geosci. J. 20(5), 681–690 (2016)
14. Taylan, O., Karagözoğlu, B.: An adaptive neuro-fuzzy model for prediction of students academic performance. Comput. Ind. Eng. 57(3), 732–741 (2009)
15. Salleh, M.N.M., Hussain, K.: A review of training methods of ANFIS for applications in business and economics. Int. J. U-And E-Serv. Sci. Technol. 9(7), 165–172 (2016)
16. Hussain, K., Salleh, M.N.M., Cheng, S., Shi, Y.: Metaheuristic research: a comprehensive survey. Artif. Intell. Rev. 5, 1–43 (2018)
17. Rini, D., Shamsuddin, S.M., Yuhaniz, S.S.: Balanced the trade-offs problem of anfis using particle swarm optimisation. TELKOMNIKA (Telecommun. Comput. Electron. Control.) 11(3), 611–616 (2013)
18. Basser, H., Karami, H., Shamshirband, S., Akib, S., Amirmojahedi, M., Ahmad, R., Jahangirzadeh, A., Javidnia, H.: Hybrid ANFIS–PSO approach for predicting optimum parameters of a protective spur dike. Appl. Soft Comput. 30, 642–649 (2015)
19. Bagheri, A., Peyhani, H.M., Akbari, M.: Financial forecasting using anfis networks with quantum-behaved particle swarm optimization. Expert Syst. Appl. 41(14), 6235–6250 (2014)
20. Liu, P., Leng, W., Fang, W.: Training ANFIS model with an improved quantum-behaved particle swarm optimization algorithm. Math. Prob. Eng. 2013, 10 (2013)
21. Nhu, H.N., Nitsuwat, S., Sodanil, M.: Prediction of stock price using an adaptive neuro-fuzzy inference system trained by firefly algorithm. In: 2013 International Computer Science and Engineering Conference (ICSEC), pp. 302–307. IEEE (2013)
22. Salleh, M.N.M., Hussain, K., Naseem, R., Uddin, J.: Optimization of ANFIS using artificial bee colony algorithm for classification of malaysian SMES. In: International Conference on Soft Computing and Data Mining, pp. 21–30. Springer (2016)
23. Karaboğa, S., Kaya, E.: Training ANFIS by using the artificial bee colony algorithm. Turk. J. Electr. Eng. Comput. Sci. 25(3), 1669–1679 (2017)
24. Orouskhani, M., Mansouri, M., Orouskhani, Y., Teshnehlab, M.: A hybrid method of modified cat swarm optimization and gradient descent algorithm for training anfis. Int. J. Comput. Intell. Appl. 12(02), 1350007 (2013)

25. da Costa Martins, J.K.E., Araújo, F.M.U.: Nonlinear system identification based on modified ANFIS. In: International Conference on Informatics in Control, Automation and Robotics (ICINCO), pp. 588–595 (2015)
26. Peymanfar, A., Khoei, A., Hadidi, Kh.: A new ANFIS based learning algorithm for CMOS neuro-fuzzy controllers. In: 2007 14th IEEE International Conference on Electronics, Circuits and Systems, ICECS 2007, pp. 890–893. IEEE (2007)
27. Şahin, M., Erol, R.: A comparative study of neural networks and anfis for forecasting attendance rate of soccer games. Math. Comput. Appl. 22(4), 43 (2017)
28. Salleh, M.N.M., Talpur, N., Hussain, K.: Adaptive neurofuzzy inference system: overview, strengths, limitations, and solutions. In: International Conference on Data Mining and Big Data, pp. 527–535. Springer (2017)
29. Karaboga, D.: An idea based on honey bee swarm for numerical optimization. Technical report, Technical report-tr06, Erciyes university, engineering faculty, computer engineering department (2005)

An Efficient Robust Hyper-Heuristic Algorithm to Clustering Problem

Mohammad Babrdel Bonab[1,2(✉)], Yong Haur Tay[1],
Siti Zaiton Mohd Hashim[2], and Khoo Thau Soon[1]

[1] Centre for Computing and Intelligent Systems (CCIS),
Universiti Tunku Abdul Rahman, Selangor, Malaysia
bbmohammad2@live.utm.my, tayyh@utar.edu.my,
khoothausoon@yahoo.com
[2] Big Data Centre & Soft Computing Research Group,
Universiti Teknologi Malaysia, Johor Bahru, Malaysia
sitizaiton@utm.my

Abstract. Designing and modeling an optimization algorithm with dedicated search is a costly process and it need a deep analysis of problem. In this regard, heuristic and hybrid of heuristic algorithms have been widely used to solve optimization problems because they have been provided efficient way to find an approximate solution but they are limited to use number of different heuristic algorithm and they are so problem-depend. Hyper-heuristic is a set of heuristics, meta- heuristics, and high-level search strategies that work on the heuristic search space instead of solution search space. Hyper-heuristics techniques have been employed to develop approaches that are more general than optimization search methods and traditional techniques. The aim of a hyperheuristic algorithms is to reduce the amount of domain knowledge by using the capabilities of high-level heuristics and the abilities of low-level heuristics simultaneously in the search strategies. In this study, an efficient robust hyperheuristic clustering algorithm is proposed to find the robust and optimum clustering results based on a set of easy-to-implement low-level heuristics. Several data sets are tested to appraise the performance of the suggested approach. Reported results illustrate that the suggested approach can provide acceptable results than the alternative methods.

1 Introduction

Clustering is an un-supervised method in the data mining and pattern recognition. Nevertheless, most of the clustering algorithms are unstable and very sensitive to their input parameters. The aim of clustering is to classify or broken data into different k classes, such that each of these cluster includes data, which has maximum dissimilarity and most similarity with the other existing clusters [1, 2]. The k-means clustering algorithm is one of basic algorithms in the literature of machine learning and pattern recognition. Yet, due to the random initialization and the adherence of outputs to initial-cluster centres, the risk of trapping into local optimum ever exists. In another words, its efficiency highly depends on the parameters of algorithm and first initial cluster centers.

© Springer Nature Switzerland AG 2019
S. Omar et al. (Eds.): CIIS 2018, AISC 888, pp. 48–60, 2019.
https://doi.org/10.1007/978-3-030-03302-6_5

The main goal of k-means algorithm is to minimize the diversity of data in a class from their cluster centre. K-means initialization problem is considered by meta-heuristic and heuristic methods, however it still risks being trapped in local optimality. There are several studies to deal with k-means problem. For example, Ishak Boushaki et al. have presented a new quantum-chaotic cuckoo-search algorithm (QCCS) based on cuckoo-search (CS) meta-heuristic algorithms to tackle with clustering problems. they improved the cuckoo search abilities by using the quantum theory in order to deal with cuckoo search clustering problem in terms of global search capability [3]. Haiyun et al. have introduced an improved pigeon-inspired optimization (IPIO) algorithm to deal with random initialization and the risk of trapping into local optimum [4]. Liping Sun et al. have proposed a hybrid data clustering algorithm based on Density peaks clustering (DPC) and Gravitational Search Algorithms (GSA) to use the use advantages of nearest neighbor method and distance measure. Sun et al. they have used the Density peaks clustering (DPC) to random initialization and Gravitational Search Algorithm (GSA) to optimize the clustering center set [5]. Nahlah M. Shatnawi has presented a hybrid method based on combination of Lévy motion or Lévy-flights and Bees algorithm to tackle the clustering problem and random initialization [6]. Bonab et al. have introduced a combination of imperialist competitive algorithm (ICA) and seed-cluster-center algorithm (SCCA) to find the best cluster center [7]. In another study, they proposed a combined algorithm based on artificial bee colony (ABC) and differential evolution (DE) to deal with random initialization in k-means algorithm [8, 9]. It has been proven that existing meta-heuristic based clustering algorithms outperform traditional clustering algorithms, but these frequently have limitations, thus resulting in the use of several combinations of algorithms. The first problem is the limitations of metaheuristic and hybrid of metaheuristic based clustering algorithms in the search for solutions within the solution space. This has made it necessary to have a hyperheuristic clustering algorithm without any limitations and with a dynamic section for the setting of parameters in order to increase the power of exploration and exploitation within the solution space. The second problem is the absence of algorithm for validating and interpreting the heuristics during the process of clustering algorithms. The main objective of the current research is to propose an efficient and robust hyperheuristic algorithm to overcome clustering problem, by using different low-level heuristics and a powerful high-level heuristic.

In the next sections of paper, in Sect. 2 the data clustering is reviewed. In Sects. 3 and 4, the k-means problem and different parts of proposed hyperheuristic algorithm for clustering problem is introduced. In Sect. 5, experimental results of proposed hyper-heuristic algorithm are compared and shown with alternative methods on real datasets and in another experimental, on the benchmark and industrial images have been applied and in the last Sect. 6 the conclusions is included.

2   Clustering

Clustering of data is one of the most widely used techniques in machine learning and data mining methods. Clustering is determined as a set of instances similar to each-other and diverse from the instances of other classes or clusters [10]. Indeed, the instances inside one class have most similarity with each other and maximum diversity with other classes of objects.

In classic description, the hard type-clustering algorithm of $S = \{s_1, s_2, \ldots, s_m\}$ is a dataset of m samples, which are to be classified or grouped in to k clusters as $C = \{C_1, C_2, \ldots, C_k\}$ such that each of these clusters satisfies the following:

1. $\cup_{i=1}^{k} C_i = S$
2. $C_i \neq \emptyset \, i = 1 \ldots k$
3. $C_i \cap C_j \neq \emptyset, \quad i, j = 1 \ldots k, \quad i \neq j$

Therefore the number of different states for grouping m samples into k cluster will be:

$$NG(m, k) = \frac{1}{k!} \times \sum_{i=0}^{k} (-1)^i \times (k - i)^m \times \binom{k}{i} \qquad (1)$$

In some of approaches, clusters number is specified manually, therefore as mentioned in Eq. 1, finding the best scenario for clustering problem is so hard, with a given k.

As well as clustering states from m samples to k clusters exponentially increases by $(k^{\wedge}m)/k!$. Given this problem, finding the best way for grouping m samples in to k clusters is an NP Complete problem which should be optimally solved by engaging some techniques.

3   K-Means Algorithm

K-means algorithm is one of the simple and fast method, which is universally applied because of its simplicity, little iteration and fast convergence. K-means clustering algorithm try to find cluster centres $(m_1, m_2, \ldots, m_k)$ by minimizing sum of squared distance of each instance x_i from the closest cluster m_j. The performance of k-means clustering algorithm is influenced by initial selection method for cluster centers. In the below, four stages are proposed for k-means clustering algorithm to cluster data.

1. Select k objects 'randomly' from $S = \{s_1, s_2, \ldots, s_k\}$ as cluster centers of $(m_1, m_2, \ldots, m_k)$
2. Add object s_i from $S = \{s_1, s_2, \ldots, s_k\}$ to cluster m_j using Eq. (2):

$$\|s_i - m_j\| < \|s_i - m_p\| \, j \neq p, \quad 1 \leq p \leq k \qquad (2)$$

3. Based on the clustering of step 2, the new cluster centres are computed based on the Eq. (3), where, n_j is the number of instances in the considered cluster:

$$m_i^* = \frac{1}{n_j} \sum_{s_j \in C_i} s_j, \quad 1 \leq i \leq k \qquad (3)$$

4. Repeat clustering algorithm from step two, if the cluster centres were changed, otherwise make the clustering according to the resulted centres.

As previously noted, k-means performance relies on initial centres, which is a major challenge in k-means clustering algorithm. Accordingly, Random selection of initial cluster centres makes k-means clustering algorithm yield out various results for different runs over the same datasets, which is intended one of potential disadvantages of k-means clustering method. By using the hyperheuristic algorithm two goals are followed, first, the high-level heuristic in hyperheuristic framework with information regarding the main problem will not work, but works on the fitness function values that are achieved form low-level heuristics. All necessary information on the problems is investigated with the low-level heuristic, this matter makes algorithm as generality and make it easy to use for other problems. Second, using simple and easy heuristics in low level of algorithm, eliminates expression about problem as specific form. This subject also make algorithm as comprehensiveness [11].

3.1 Fitness Function

In this case, the distance between the clusters centres and each instances, for calculating the fitness of each solution will be used. Accordingly, firstly a set of cluster centers will be selected randomly and then clustering of the numerator will be computed according to Eq. (2). Now based on the centres obtained in the interaction stage, fitness of solutions and the new cluster centers based on Eq. (3) will be computed.

$$\text{Fitness}(M) = \sum_{i=1}^{k} \sum_{s_j \in C_i} \left\| s_j - m_i^* \right\| \tag{4}$$

4 Proposed Hyper-Heuristic Algorithm

A brief observation on clustering problem studies in recent years, indicated that the vast majority of the used approaches to deal with this problem were using metaheuristic and heuristics and most of them had specific knowledge of problems to improve or build solutions. In the past few years, research on hyperheuristic is focused on increasing the generality of optimization systems. This study is considered an investigation of component-based hyperheuristics. Figure 1 shows different parts of a hyperheuristic algorithm. The proposed algorithm includes two main parts as high-level and low-level heuristics. In high-level heuristic the different methods and mechanism can be used to generate and choose low-level heuristics. In this study, we have used the hybrid of evolutionary algorithm (SAGA) which use a population-based Simulated Annealing algorithm with mutation and crossover operations (Genetic algorithm operations). The main advantage of population-based simulated annealing algorithm is its capability in discrete domains to select different low-level heuristics based on their performances. This helps us to generate new combination of different heuristics based on the existing heuristics. In the low-level heuristic part, we have used a pool of heuristic algorithm including different heuristics including classic heuristic, meta-heuristic and their operations. Using different heuristic algorithms have significant benefits over mathematical techniques.

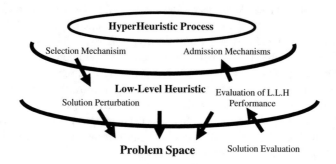

Fig. 1. HyperHeuristic algorithm

The descriptions of the proposed algorithm below is given and shown in Fig. 2. Hyperheuristics is a relatively new field, which has proven to be effective for combinatorial optimization problems to offer generalized solutions. Evolutionary based approaches have played real role in the advancement of hyperheuristics algorithms. Evolutionary based hyperheuristics were successfully used in various domains to solve problems including vehicle routing, financial forecasting, permutation flow-shop, packing problems and educational timetabling. Our proposed clustering algorithm has managed to minimize the dissimilarity of all points of a cluster using hyperheuristic method, from the gravity center of the cluster with respect to capacity constraints in each cluster. The algorithm of hyperheuristic has emerged from pool of heuristic techniques. Mapping between solution spaces is one of the powerful and prevalent techniques in optimization domains. Most of the existing algorithms work directly with solution spaces where in some cases is very difficult and is sometime impossible due to the dynamic behavior of data and algorithm. By mapping the heuristic space into solution spaces, it would be possible to make easy decision to solve clustering problems. The suggested method solves the getting trapped in local optimality and sensitivity problem of initial values by applying the hyperheuristic clustering algorithm to cluster data with different low-level heuristics.

4.1 Simulated Annealing Combined with Genetic Operations (SAGA)

The SAGA approach is especially developed for drawback of simulated annealing, where simulated annealing works only with one solution and in order to increase efficiency and diversity of the suggested algorithm have been used population based simulated annealing instead of working with one solution in simulated annealing (SA) algorithm. In this part as high-level heuristic, to some extent genetic algorithm and population based simulated annealing have been used. In the first stage for initialization, the heuristics are evaluated based on their merit and all their individual attributes. According to their merits with proposed SAGA, heuristics are selected to be used in problem space to select the best solution, because the proposed SAGA algorithm is no sensitive to initial value or specific heuristic. In this algorithm, for all iteration have been considered a population where, Ne is number of neighbors for each population that in this manner each population is generated Ne neighbors around every solutions and nP is number of population, which altogether are equal with $Ne \times nP$. The pseudo-code of the suggested SAGA algorithm illustrated in Algorithm 1.

Algorithm 1: Pseudo code of the '*SAGA* algorithm'

---

1. **Inputs** : Data set I= $\{x_1, x_2, \dots, x_m\}$), Heuristic pool and Heuristic information;
2. **Output** : selected heuristics (H = $\{h_1, h_2, \dots, h_n\}$)
3. **Begin**
5. Generate-Initial-parameters ();   // Primary evaluation and production
6. Initial-Evaluation-on-Population ();   // specify the best response found
7. Tep = Tep0;   // Configuration of initial temperature
8. MaxIter= Number_of_Iterations;   // Iterations
9. **While (it<MaxIter)**
  10. For each member of the population a certain number of neighbors is produced and marked.
  11. Create Mutation and Crossover populations to create neighbors for each solutions
  12. Members of the mutation and crossover population with neighbors population sorted and among them best members identified in a number of the main population.
  13. Each member of the main population with one of the members of winner neighbors compared according to the SA.
  14. The best list of heuristic found upgraded (each new solution produced is compared with the previous record);
  15. Temperature should be reduced and steps resumed from step 9.
  16. **End**

4.2   Heuristic Selection Mechanism

Selection Mechanism is one of the important stage in the high-level algorithm where heuristics are selected between low-level heuristics from a pool of heuristics for later processes. There were several methods for selection that could be used to choose a set of low-level heuristics. The intelligent selection system is applied to select low-level heuristics under managed selection. This system is a high-level mechanism to specify the suitable low-level heuristics [12]. There are several heuristic selection methods but in this study, three of these methods are used. This mechanism specified the selection probability of each low-level heuristic. Fitness proportionate selection or Roulette Wheel Selection (RWS), Tournament Selection, and Random Selection (RND) were used to select the low-level heuristics in this study.

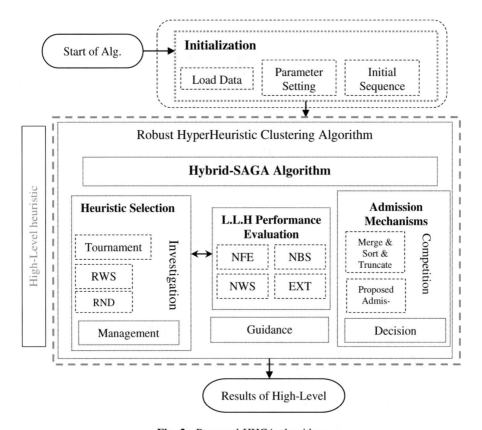

Fig. 2. Proposed HHCA algorithm

4.3 Low-Level Heuristics Performance Evaluation

The fitness of each chromosome in the population is computed by using the chromosome to create new heuristics. For example, consider the element 12443. The first selection would be generated and selected randomly. The fitness measure of the chromosome is a function that is calculated based on the Number of Function Evaluation (NFE), Number of new Best Solutions found (NBS), Number of Worst Solutions found (NWS or Non-NSB) and Execution Time (EXT) (Fig. 3).

4.4 Admission Mechanisms

To specify the diversification and intensification of the hyperheuristic behavior, the admission mechanism is used to increase the performance of the algorithm. To use this mechanism is to use the threshold value as an acceptance mechanism. In this study, the acceptance criterion is used from [13]. Simulated annealing and random selection are two admission mechanism criteria. Like the heuristic selection section, the admission mechanism framework had two main systems, which were the learning and decision system. Acceptance of simulated annealing worked based on improving solutions and

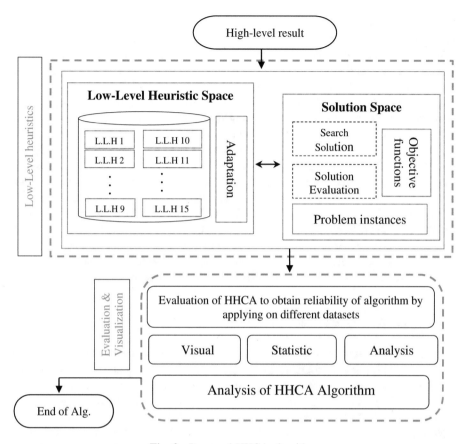

Fig. 3. Proposed HHCA algorithm

random selection for no improving solutions. Simulated annealing used the intensification and diversification simultaneously. Bad solutions are also selected if their fitness quality are less than the threshold value. Random selection acceptance is the second admission criteria that chose solutions as randomly. The motivation behind this mechanism is to measure the quality of the heuristics based on their searching operation performance. In this method, new solutions that are created under the proposed algorithm are more than the limit-size and needed to be elected for the next iteration. According to this scenario, all the solutions that are obtained by discussed methods are combined and then sorted with together. Finally, the best solutions for next generation equal to the original-population are elected and other solutions are truncated.

4.5 Low-Level Heuristics

One of the motivations of hyperheuristics was to take advantage of different heuristics and algorithms. In most of the heuristic algorithms, when these heuristics were applied separately, it seemed impossible to find solutions with similar qualities to what could be obtained when all the heuristics were combined. Therefore, using several algorithms together would lead to better solutions than applying each one individually. Unfortunately, hybrid algorithms and combined methods were limited to multiple algorithms. Hyperheuristic algorithms dealt with these kinds of problems by taking advantage of the strengths of the different heuristics.

A set of heuristic and meta-heuristic algorithms were considered as low-level heuristics. In order to handle the clustering problems and find the best cluster center with minimum distance between data and the centers, seven low level heuristics were considered. The following low-level heuristics, based on crossover, mutation, and moving operations, were used as a low-level hyperheuristic set. In this thesis, we used two kinds of low-level heuristics which included 10 crossover and 5 mutation kinds (Arithmetical Crossover, Average Crossover, Blend Crossover, Discrete Crossover, Double Point Crossover, Flat Crossover, Heuristic Crossover/Intermediate Crossover, Shuffle Crossover, Single Point Crossover, and Uniform Crossover). These are available low-level heuristic in clustering problem to deal with trapping to local optimum problem and local search heuristics to find the best solutions in continue domain problem, this is one of the motivation to select these low-level heuristics.

5 Experimental Results

The goal of this study was to access the optimum cluster centers based on proposed hyperheuristic framework for data clustering and compare it with state-of-the-art hybrid clustering approaches. To achieve this, we considered ten other algorithms, which included: (1) hybrid of PSO and ACO, called PSO – ACO, (2) Particle Swarm Optimization - PSO, (3) Simulated Annealing – SA, (4) Tabu search – TS, (5) Genetic Algorithm – GA, (6) Ant Colony Optimization – ACO, (7) Honey-Bees Mating Optimization – HBMO, (8) hybrid of PSO and SA, called PSO_SA, (9) hybrid of ACO and SA called ACO – SA and (10) K-Means algorithm. Methods (1, 8, and 9) are hybrid algorithms and the rest are single evolutionary algorithms each methods has their own different advantages and disadvantages which are reported in reference [14] in the Tables 2, 3, 4, 5, 6 and 7.

To evaluate the accuracy and efficiency of the hyperheuristic algorithm, experiments were performed on six standard datasets, and image datasets to determine the correctness of clustering algorithms. The suggested algorithm was coded in an appropriate programming language (MATLAB) which ran on a computer with a 3.60 GHz microprocessor and 4 GB main-memory. To measure the efficiency of the suggested algorithm, the standard-data items of below table were used (Table 1).

5.1 Evaluating Standard Datasets Based on Proposed Algorithm

To better analysis and study of the suggested method, the execution results of the proposed approach along with PSO-ACO-K, ACO-SA, PSO-SA, HBMO, ACO, PSO, SA, GA, K-Means and TS results which are introduced in Ref [14] are tabulated in the Tables 2, 3, 4, 5, 6 and 7. In mentioned tables worst, best and average results are presented respectively for 100-runs. The resulted tables shows the distance between

Table 1. Table Type-Styles

Names of dataset	Data set attribute		
	Size of data set	No. of clusters	No. of attributes
Glass	214(70,217,76,13,9,29)	6	9
Vowel	871(72,89,172,151,207,180)	6	3
CMC	1473 (629, 333, 511)	3	9
Iris	150(50,50,50)	3	4
Cancer	683(444,239)	2	9
Wine	178 (59,71,48)	3	13

cluster centers and each data object, which is mentioned in Sect. 3.1. As simply observed in the reported results, the suggested methods generates acceptable solutions based on the execution time.

Table 2. The results over IRIS dataset for 100 Runs

Method	Result			CPU	STD
	Worst	Average	Best	Time(S)	
PSO–ACO	96.674	96.654	96.654	~17	0.17
PSO	97.897	97.232	96.8942	~30	0.57
SA	102.01	99.957	97.457	~32	0.98
TS	98.569	97.868	97.365	~135	13.84
GA	139.778	125.197	113.986	~140	4.62
ACO	97.808	97.171	97.1	~75	0.29
HBMO	97.757	96.953	96.752	~82	0.18
PSO_SA	96.678	96.67	96.66	~17	0.65
ACO–SA	96.863	96.731	96.66	~25	0.19
K-Means	120.45	106.05	97.333	0.4	85.5
HHCA Alg.	**96.5403**	**96.5403**	**96.5403**	**~14**	**0**

Table 3. The results over CMC dataset for 100 Runs

Method	Result			CPU	STD
	Worst	Average	Best	Time(S)	
PSO–ACO	5,697.42	5,694.92	5,694.51	~135	0.868771
PSO	5,923.24	5,820.96	5,700.98	~131	46.95969
SA	5,966.94	5,893.48	5,849.03	~150	50.8672
TS	5,999.80	5,993.59	5,885.06	~155	40.84568
GA	5,812.64	5,756.59	5,705.63	~160	50.3694
ACO	5,912.43	5,819.13	5,701.92	~127	45.6347
HBMO	5,725.35	5,713.98	5,699.26	~123	12.69
PSO_SA	5,701.81	5,698.69	5,696.05	~73	1.8691
ACO–SA	5,700.26	5,698.26	5,696.60	~89	1.98238
K-Means	5,934.43	5,893.60	5,842.20	0.5	47.16
HHCA Alg.	**5,532.18**	**5,532.18**	**5,532.18**	**~30**	**0**

Table 4. The results over GLASS dataset For 100 Runs

Method	Result			CPU	STD
	Worst	Average	Best	Time(S)	
PSO–ACO	200.01	199.61	199.57	~35	0.13914
PSO	283.52	275.71	270.57	~400	4.557134
SA	287.18	282.19	275.16	~410	4.238458
TS	286.47	283.79	279.87	~410	4.192734
GA	286.77	282.32	278.37	~410	4.138712
ACO	280.08	273.46	269.72	~395	3.584829
HBMO	249.54	247.71	245.73	~390	2.43812
PSO_SA	202.45	201.45	200.14	~38	0.89243
ACO–SA	202.76	201.89	200.71	~49	0.887234
K-Means	255.38	235.5	215.74	~1	12.47107
HHCA Alg.	**210.46**	**210.43**	**210.42**	**~27**	**0.0078305**

Table 5. The results over WINE dataset for 100 Runs

Method	Result			CPU	STD
	Worst	Average	Best	Time(S)	
PSO–ACO	16,297.93	16,295.92	16,295.34	~33	0.868771
PSO	16,562.31	16,417.47	16,345.96	~123	85.4974
SA	18,083.25	17,521.09	16,473.48	~129	753.084
TS	16,837.53	16,785.45	16,666.22	~140	52.073
GA	16,530.53	16,530.53	16,530.53	~170	0
ACO	16,530.53	16,530.53	16,530.53	~121	0
HBMO	16,357.28	16,357.28	16,357.28	~40	0
PSO–SA	16,296.10	16,296.00	16,295.86	~38	0.89612
ACO–SA	16,322.43	16,310.28	16,298.62	~84	10.62197
K-Means	18,563.12	18,061.01	16,555.68	0.7	793.213
HHCA Alg.	**16,292.18**	**16,292.18**	**16,292.18**	**~24**	**0**

Table 6. The results over VOWEL dataset for 100 Runs

Method	Result			CPU	STD
	Worst	Average	Best	Time(S)	
PSO–ACO	1,49,101.68	1,48,995.20	1,48,976.01	~17	24.5420931
PSO	1,49,121.18	1,48,999.83	1,48,976.02	~30	28.8134692
SA	1,65,986.42	1,61,566.28	1,49,370.47	~32	2,847.09
TS	1,65,996.43	1,62,108.54	1,49,468.27	~33	2,846.24
GA	1,65,991.65	1,59,153.50	1,49,513.74	~39	3,105.54
ACO	1,65,939.83	1,59,458.14	1,49,395.60	~27	3,485.38
HBMO	1,65,804.67	1,61,431.04	1,49,201.63	~32	2,746.04
PSO_SA	1,54,751.26	1,50,343.45	1,49,001.85	~33	351.45
ACO–SA	1,49,364.30	1,49,141.40	1,49,005.00	~35	120.38
K-Means	1,61,236.81	1,59,242.89	1,49,422.26	~0.45	916
HHCA Alg.	**1,49,087.39**	**1,49,005.39**	**1,48,967.24**	**~13**	**38.7983**

Table 7. The results over CANCER dataset for 100 Runs

Method	Result			CPU	STD
	Worst	Average	Best	Time(S)	
PSO–ACO	2,964.50	2,964.39	2,964.38	~17	0.037947
PSO	3,318.88	3,050.04	2,973.50	~123	110.8013
SA	3,421.95	3,239.17	2,993.45	~126	230.192
TS	3,434.16	3,251.37	2,982.84	~130	232.217
GA	3,427.43	3,249.46	2,999.32	~135	229.734
ACO	3,242.01	3,046.06	2,970.49	~123	90.50028
HBMO	3,210.78	3,112.42	2,989.94	~136	103.471
PSO_SA	2,967.41	2,966.32	2,965.17	~28	1.7201
ACO–SA	2,968.29	2,966.63	2,967.83	~40	1.7732
K-Means	3,521.59	3,251.21	2,999.19	~0.5	251.14
HHCA Alg.	**2,964.38**	**2,964.38**	**2,964.38**	**~22**	**0**

According to reported results in Tables 2, 3, 4, 5, 6 and 7 the proposed hyper-heuristic algorithm over WINE, CMC, IRIS, VOWEL, WINE, GLASS, and CANCER Datasets provides the best and acceptable outputs in comparison with other mentioned methods. The standard deviation of the fitness function for hyper-heuristic algorithm (HHCA) is zero. This means that proposed hyper-heuristic clustering algorithm (HHCA) is very reliable and precise. In other words, it provides the small standard deviation and optimum value in comparison to those of other algorithms. The results obtained on the different datasets show that hyper-heuristic clustering algorithm (HHCA), most time converges to the best and acceptable local optimum. The standard deviation of the cost function for proposed algorithm in the most time, is zero, which is significantly less than other methods.

5.2 Image Segmentation Using Proposed Algorithm

Digital-Image segmentation is a challenging and complex issues, which can be different depending on the nature of inherent of pictures. There are several approaches for image segmentation challenges. In the previous section, it was shown that the suggested hyperheuristic method was one of best approaches for clustering of data. To further test the performance of the method, it was tested on one industrial image and one standard image. In recent years, most of researches have been done on the evolutionary-based algorithm inspired from human societies and natures. Figures 4 and 5 shows, image samples for this study. For finding the best cluster centers and optimum cluster centers the proposed HHCA approach is used to find minimum distance between each cluster centers with each data object. Proposed HHCA algorithm should be able to separate the segments of different pictures in different light condition with high precision and accuracy.

By specifying, the peaks locations in the image histograms can obtain the optimal threshold for image segmentation According to the reported results in Fig. 5, the suggested algorithm over different images with different conditions provides the

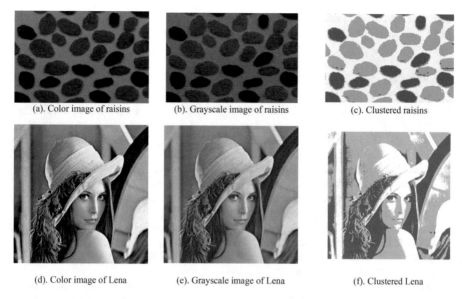

(a). Color image of raisins   (b). Grayscale image of raisins   (c). Clustered raisins

(d). Color image of Lena   (e). Grayscale image of Lena   (f). Clustered Lena

Fig. 4. Used images for segmentation with proposed algorithm

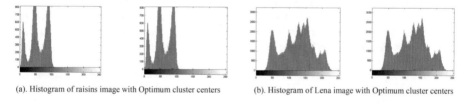

(a). Histogram of raisins image with Optimum cluster centers   (b). Histogram of Lena image with Optimum cluster centers

Fig. 5. Used images for image segmentation in Grayscale mode

acceptable segmentations and no depend on behavior and situations of data. It can be work on the different behavior and the different situations.

6   Conclusions

The main goal of this study was to propose a new robust hyperheuristic clustering algorithm that could produce efficient and high quality performance across various low-level heuristic sets in solving generic clustering problems to minimize the dissimilarity of all objects of a cluster. In this study, hyperheuristic clustering algorithm with an effective high-level and low-level heuristics algorithm were used for optimal solutions of objects in images and datasets. To increase the exploitation of hyperheuristic algorithm the different low-level hyperheuristic are used. For hyperheuristic algorithms, problem-dependent knowledge is not required. Therefore, it is easily applicable for any kind of problems that need searching. For this reason, hyperheuristic algorithms seek heuristic spaces instead of solution spaces.

Acknowledgment. The authors would like to express their cordial thanks to Universiti Tunku Abdul Rahman (UTAR) for research university grant with number of (4461/002).

References

1. Gan, G., Ma, C., Wu, J.: Data clustering: theory, algorithms, and applications, vol. 20. Siam (2007)
2. Karaboga, D., Basturk, B.: A powerful and efficient algorithm for numerical function optimization: artificial bee colony (ABC) algorithm. J. Global Optim. **39**(3), 459–471 (2007)
3. Boushaki, S.I., Kamel, N., Bendjeghaba, O.: A new quantum chaotic cuckoo search algorithm for data clustering. Expert Syst. Appl. **96**, 358–372 (2018)
4. Li, H., et al.: An improved pigeon-inspired optimization for clustering analysis problems. Int. J. Comput. Intell. Appl. **16**(02), 1750014 (2017)
5. Sun, L., et al.: An Optimized Clustering Method with Improved Cluster Center for Social Network Based on Gravitational Search Algorithm. Springer, Cham (2017)
6. Shatnawi, N.M.: Data clustering using Lévy flight and local memory bees algorithm. Int. J. Bus. Intell. Data Min. **12**(1), 14–24 (2017)
7. Babrdelbonb, M., Hashim, S.Z.M.H.M., Bazin, N.E.N.: Data analysis by combining the modified k-means and imperialist competitive algorithm. Jurnal Teknologi **70**(5) (2014)
8. Bonab, M.B., Hashim, S.Z.M.: Image segmentation with genetic clustering using weighted combination of particle swarm optimization. In: 14th International Conference on Applied Computer and Applied Computational Science (ACACOS 2015) (2015)
9. Bonab, M., et al.: Modified K-means combined with artificial bee colony algorithm and differential evolution for color image segmentation. In: Phon-Amnuaisuk, S., Au, T.W. (eds.) Computational Intelligence in Information Systems, pp. 221–231. Springer, Cham (2015)
10. Bonab, M.B., Mohd Hashim, S.Z.: Improved k-means clustering with Harmonic-Bee algorithms. In: Fourth World Congress on Information and Communication Technologies (WICT) (2014)
11. Kao, Y.-T., Zahara, E., Kao, I.W.: A hybridized approach to data clustering. Expert Syst. Appl. **34**(3), 1754–1762 (2008)
12. Mısır, M., et al.: An Intelligent Hyper-Heuristic Framework for CHeSC 2011. In: Hamadi, Y., Schoenauer, M. (eds.) Learning and Intelligent Optimization, pp. 461–466. Springer, Heidelberg (2012)
13. Mısır, M., et al.: A new hyper-heuristic as a general problem solver: an implementation in HyFlex. J. Sched. **16**(3), 291–311 (2013)
14. Niknam, T., Amiri, B.: An efficient hybrid approach based on PSO, ACO and k-means for cluster analysis. Appl. Soft Comput. **10**(1), 183–197 (2010)

Test Case and Requirement Selection Using Rough Set Theory and Conditional Entropy

Noor Fardzilawati Md Nasir[✉], Noraini Ibrahim[✉],
Mustafa Mat Deris[✉], and Mohd. Zainuri Saringat[✉]

Faculty of Computer Science and Information Technology,
Universiti Tun Hussein Onn Malaysia, 86400 Batu Pahat, Johor, Malaysia
fasha_nasir@yahoo.com.sg,
{noraini,mmustafa,zainuri}@uthm.edu.my

Abstract. The growing size and complexity of the software system makes testing essential in software engineering. In particular, the effectiveness of generating test cases becomes a crucial task, where there is an increment of source codes and a rapid change of the requirements. Therefore, the selection of effective test cases becomes problematic, when the test cases are redundant and having common requirements. Thus, new challenges arose to reduce the unnecessary test cases and find common requirements that would increase the cost and maintenance of the software testing process. To address this issue, this study proposed a technique that minimized the test cases and requirement attributes, without compromising on fault detection capability. The proposed technique, using *Rough Set Theory-Similarity Relation*, was used to reduce the size of the test cases. Subsequently, a new approach, known as *Conditional Entropy-Based Similarity Measure,* was introduced to obtain a minimum subset of requirements. It was anticipated that, the technique applied would contribute significantly towards solving the testing problems, since testers would no longer be required to select an arbitrary test suite on test runs. The proposed technique was found to have achieved up to 50% reduction of the processing time, as compared with the base-line techniques, such as, MFTS Algorithm, FLOWER, RZOLTAR and Weighted Greedy Algorithm.

Keywords: Test case reduction · Similarity relation · Conditional entropy

1 Introduction

In software testing, test case generation is the most challenging step, compared with the other software testing part [1, 2]. As the software development embarks on a new height, due to the integration of the system, the size of the test suite increased accordingly, to maintain the system. To enhance the generation of test cases, many automation techniques have been introduced to replace the manual approach, forcing for new test cases to be added to the test suites to cater for the changes in the software. Due to the many versions of the development, the possibility of redundant test cases being generated in the test suites is higher [1]. Additionally, the redundant test cases must exercise software requirements for which they were generated. Some

© Springer Nature Switzerland AG 2019
S. Omar et al. (Eds.): CIIS 2018, AISC 888, pp. 61–71, 2019.
https://doi.org/10.1007/978-3-030-03302-6_6

requirements are also common, with respect to any of the test cases. All these issues create the motivation for creating a good technique on selecting the minimal subset of test cases that covers all requirements, without hampering the decision of pass or fail of the system.

Many researchers have proposed Test Case Reduction techniques to find a minimal subset of test cases. Various techniques are applied, aimed to reduce the number of test cases, but with less attention given to maintain the fault detection capability. Hence, there is a need for a technique that produces the minimal subset of test cases, while simultaneously maintaining the ability to detect faults. Maung and Win [3] suggested entropy gain theory to get the best test suite that covers all user access in web application, where an entropy value analysis was used to signify that, test cases had met requirements, as in the original test suite, without over reducing user session data. Nevertheless, the technique requires a prior knowledge of the mathematical models to reduce generated redundant test cases [4]. At the same time, with the entropy technique, test cases and exercised requirements could be classified into a smaller size, thus reducing the computational time to run the test.

Rough Set Theory has been used for attribute selection in Incomplete Information Systems with significant successes [5, 6] with a minimal reduction in incomplete decision system. Thus, capitalizing on its advantage in handling flexibility and precise data classification in Information System (IS), Rough Set Theory was utilized for Software Testing (ST). While IS focusing on finding the attribute which is low in similarity and certainty relation, ST needs a minimal similarity relation among the test cases to minimize the redundancy between the test cases. This gave further motivation for the researcher, as it seemed to be a very attractive solution for the test case reduction issues. This method could be more meaningful, since it was combined with the second approach, which was Conditional Entropy. The application of entropy in information theory became a new heuristic for optimization and reduction of redundancy [3]. This approach attempted to get the desired common requirements exercised by the test cases to avoid a longer processing time.

The integration of Rough Set Theory and Information Theory into ST must be a practical approach in testing a real system. However, this approach is less used by researchers in ST, especially in test cases reduction. Further researches have to be done to provide better results in test cases reduction. The rest of the paper is organized as follows. In Sect. 2, the related test cases reduction works are discussed. Section 3 discusses the proposed test cases reduction in software testing. In Sect. 4, the evaluation results are presented and discussed.

2 Related Works

Xu et al. [7] proposed Weighted Greedy Algorithm for test case reduction. The algorithm starts with the determination of test cases which satisfy all the requirements, using a weighted set covering technique. This study attempted to eliminate redundant test cases and prioritized the test suite, according to the cost. The selections of essential test cases to be added in the reduced set relied on the decreasing order of priority and were done repeatedly, until all the requirements were satisfied. The results obtained showed that, the algorithm was successful to reduce 33.33% of the cost and size of test suite, with a higher efficiency.

A heuristic approach, known as RZOLTAR and proposed by Campos and Abreu [8], used the coverage matrix through the relations between test cases and the testing requirements. Typically, the matrix is mapped into a set of constraints where it evaluates a collection of optimal minimal representative set. The approach starts with the execution of system under test (SUT), using the current test suite, to obtain the coverage matrix, which in turn, is transformed into a set of constraints. The constraints are subsequently solved with MINION (off-the-shelf constraint solver) and prioritized, using a particular criterion. This approach is based on constraint solving programming, which could efficiently reduce 64.88% of the test suite size, while maintaining full coverage and fault detection capability.

Gotlieb and Marijan [9] introduced a new approach of test case reduction, known as, FLOWER, based on a search among network maximum flows. Given a Test Suite T and a set of test requirements R, FLOWER forms a flow network through the given information about test suite and requirements covered by the suite. This flow network traverses to find its maximum flows. Subsequently, Gotlieb and Marijan [9] optimally used the Ford-Fulkerson method to evaluate the maximum flows and Constraint Programming Technique to search optimal flows. The approach starts with encoding Test Suite Reduction with a Flow Network using a bipartite graph, before a Representative Test Suite using Maximum Flows is found. Finally, a Minimal-Cardinality Subset of T that covers R is found. They evaluated their work on a set of 2005 test suite reduction problem, and obtained the same reduction rate as ILP, since both approaches computed optimal solutions. However, it took 30% more time (average) and generated 5%–15% test suite reduction, compared with the simple Greedy approach.

Harris and Raju [10] proposed Maximal Frequent Test Set (MFTS) by combining the existing HSG by Harrold et al. [11] with Bi Objective Greedy Algorithm (BOG) by Parsa and Khalilian [12]. Their work optimized the test suite, based on the related testing objectives. Initially, the requirement matrix (the map of the requirements in the row with the test cases in the column) is constructed. The association between test cases and requirement attributes is represented by 0 and 1. The reduced test case was later generated, using a mathematical operation, and the results thus obtained showed that, the algorithm reduced 76.87% of the test case size and offered a high percentage of requirement coverage. It is found that, RZOLTAR [8] and MFTS [10] were the better options, compared with the other techniques, in the test suite reduction performance, with no loss of fault detection capability and a reduced processing time. However, the number of requirement attributes minimized was not mentioned, causing the same size of dataset ($S_{dataset}$) to be produced. Meanwhile, the proposed techniques will reduce the test cases and requirements attribute with the same fault detection capability as the original test suite, which resulted in a reduced size of dataset ($S_{dataset}$) and a shortened processing time.

3 Test Case and Requirements (TCR) Reduction

A complete TCR, is 4-tuple (quadruple); $TCR = (U, R, V, f)$, where $U = \{tc_1, \ldots\ldots, tc_n\}$ denotes a non-empty finite set of test cases (objects) and $R = \{r_1, \ldots\ldots, r_m\}$ denotes a finite set of requirements (attributes), $V = \cup_{r \in R} V_r$, V_r is the domain (value set) of attribute r, $f : U \times R \to V$ is a TCR function, such that, $f(tc, r) \in V_r$, for every $(tc, r) \in U \times R$, called TCR function.

3.1 Similarity Relation

Given a complete $TCR = (U, R, V, f)$, where, $R = R * \cup \{d\}$, $R*$ is a set of condition attributes and d the decision attribute, such that, $f : U \times R \to V$, for any $r \in R$, where V_r is called domain of an attribute r. In TCR, for any subset $B \subseteq R*$, the similarity relation, S, is defined by the following definition.

Definition 1. Let $TCR = (U, R, V, f)$ be a complete TCR. A similarity relation, S, is defined as

$$\forall_{tc_m, tc_n \in U} \quad S(tc_m, tc_n) \Leftrightarrow \forall_{r_j \in B} \big((r_j(tc_m) = r_j(tc_n) \big)$$

Thus,

$$S = \big\{ (tc_m, tc_n)\, tc_m \in U \wedge tc_n \in U \wedge \forall r_j \big(r_j \in B \to \big(r_j(tc_m) = r_j(tc_n) \big) \big) \big\}$$

From Definition 1, the notion of similarity class is described as follows.

Definition 2. Let $TCR = (U, R, V, f)$ be a complete TCR. The similar class $I_B^S(tc_m)$ of a test case, with reference to requirements set, B, is defined as $I_B^S(tc_m) = \{tc_n | tc_n \in U\}$.

3.2 Test Case Reduction Using Similarity Relation

Two test cases were considered indiscernible, in terms of the value of requirements, R, exercised by each test case. Thus, if $(tc_m, tc_n) \in S(R)$, then tc_m and tc_n were similar by attributes with R.

Example 1. Table 1 is a complete TCR, where $tc_1, tc_2.., tc_{12}$ are the tests cases. $r_1, r_2, \ldots r_{12}$ are the twelve requirements. The d is a decision attribute, where its domain value is {Pass, Fail}, Pass = {tc2, tc3, tc4, tc6, tc7, tc8, tc9, tc10, tc11, tc12} and Fail = {tc1, tc5}. For tc_i, $i = 1, 2, \ldots\ldots, 12$ and condition requirements attributes, $R = \{r_1, r_2, \ldots, r_{12}\}$. From Definition 1, the results can be easily obtained by analyzing it with similarity relation in Definition 2, as follows:

Table 1. A complete test case requirement matrix table, *TCR*

tc	r_1	r_2	r_3	r_4	r_5	r_6	r_7	r_8	r_9	r_{10}	r_{11}	r_{12}	Decision
tc_1	1	1	1	1	1	1	0	0	0	0	0	1	Fail
tc_2	1	1	1	1	1	1	0	0	0	0	0	1	Pass
tc_3	1	1	1	1	1	1	0	0	0	0	0	1	Pass
tc_4	1	1	1	1	1	1	0	0	0	0	0	1	Pass
tc_5	1	1	1	1	1	1	0	0	0	0	0	1	Fail
tc_6	1	1	1	1	1	1	0	0	0	0	0	1	Pass
tc_7	1	0	1	1	0	0	1	0	0	0	0	1	Pass
tc_8	1	1	1	1	1	1	0	1	0	0	0	1	Pass
tc_9	1	1	1	1	1	1	0	0	1	0	0	1	Pass
tc_{10}	1	0	1	1	1	0	0	0	0	1	0	1	Pass
tc_{11}	1	1	1	1	1	1	0	0	0	0	1	1	Pass
tc_{12}	1	1	1	1	1	1	0	0	0	0	0	1	Pass

$$I_R^S(tc_1) = \{tc_1, tc_2, tc_3, tc_4, tc_5, tc_6, tc_{12}\}, I_R^S(tc_2) = \{tc_1, tc_2, tc_3, tc_4, tc_5, tc_6, tc_{12}\},$$
$$I_R^S(tc_3) = \{tc_1, tc_2, tc_3, tc_4, tc_5, tc_6, tc_{12}\}, I_R^S(tc_4) = \{tc_1, tc_2, tc_3, tc_4, tc_5, tc_6, tc_{12}\},$$
$$I_R^S(tc_5) = \{tc_1, tc_2, tc_3, tc_4, tc_5, tc_6, tc_{12}\}, I_R^S(tc_6) = \{tc_1, tc_2, tc_3, tc_4, tc_5, tc_6, tc_{12}\},$$
$$I_R^S(tc_7) = \{tc_7\}, I_R^S(tc_8) = \{tc_8\}, I_C^S(tc_9) = \{tc_9\},$$
$$I_C^S(tc_{10}) = \{tc_{10}\}, I_C^S(tc_{11}) = \{tc_{12}\}, I_R^S(tc_{12}) = \{tc_1, tc_2, tc_3, tc_4, tc_5, tc_6, tc_{12}\}$$

and

$$I_{R^* \cup \{d\}}^S(tc_1) = \{tc_1, tc_5\}, I_{R^* \cup \{d\}}^S(tc_2) = \{tc_2, tc_3, tc_4, tc_6, tc_{12}\}, I_{R^* \cup \{d\}}^S(tc_3) = \{tc_2, tc_3, tc_4, tc_6, tc_{12}\},$$
$$I_{R^* \cup \{d\}}^S(tc_4) = \{tc_2, tc_3, tc_4, tc_6, tc_{12}\}, I_{R^* \cup \{d\}}^S(tc_5) = \{tc_1, tc_5\}, I_{R^* \cup \{d\}}^S(tc_6) = \{tc_2, tc_3, tc_4, tc_6, tc_{12}\},$$
$$I_{R^* \cup \{d\}}^S(tc_7) = \{tc_7\}, I_{R^* \cup \{d\}}^S(tc_8) = \{tc_8\}, I_{R^* \cup \{d\}}^S(tc_9) = \{tc_9\}, I_{R^* \cup \{d\}}^S(tc_{10}) = \{tc_{10}\}, I_{R^* \cup \{d\}}^S(tc_{11}) = \{tc_{11}\},$$
$$I_{R^* \cup \{d\}}^S(tc_{12}) = \{tc_2, tc_3, tc_4, tc_6, tc_{12}\}$$

Thus,

$$|S| = \left| \{ \{tc_1, tc_5\}, \{tc_2, tc_3, tc_4, tc_6, tc_{12}\}, \{tc_7\}, \{tc_8\}, \{tc_9\}, \{tc_{10}\}, \{tc_{11}\} \} \right|$$
$$= 7$$

From the above result, there were indiscernible test cases in the similar relation, where $tc_1 = tc_5$ and $tc_2 = tc_3 = tc_4 = tc_6 = tc_{12}$. As such, the 12 objects could easily be reduced to 7 different objects.

3.3   Requirement Minimization Using Conditional Entropy

Conditional entropy on the similarity relation approach was introduced to reduce a certain number of the common requirement attributes. The algorithm imposed in this research was based on breadth-first search to find the minimal requirement selection for TCR. The algorithm was given as in Algorithm 1.

Input: A complete TCR, $TCR = (U, R, V, f)$

Output: A selection requirement, SR

a) For every size $= 0$ to $|R^*|$

b) For all subset selectRequirementAttribute

with selectRequirementAttribute =size

c) If $EN(d|\text{selectRequirementAttribute}) \neq EN(d|R^*)$, go to step b, otherwise return $M = \text{selectRequirementAttribute}$

d) End

e) End

Algorithm 1: Breath-first search for requirement selection

Definition   3. Given   TCR, $TCR = (U, R^* \cup \{d\})$ and $B \subseteq R^*$.   Let $U/S_{R^*} = \{S_{R^*}(tc_1), S_{R^*}(tc_2), \ldots, S_{R^*}(tc_{|U|})\}$, $U/d = \{d_1, d_2, \ldots, d_k\}$. The conditional entropy of R^* to d is defined as follows [6]:

$$EN(d \mid R^*)$$

$$= -\sum_{i=1}^{|U|} p(S_{R^*}(tc_i))$$

$$\times \sum_{j=1}^{|U/d|} p\left(d_j \mid S_{R^*}(tc_i)\right) \, \log_2 p\left(d_j \mid S_{R^*}(tc_i)\right)$$

Where, $p\left(S_{R^*}(tc_i)\right) = \frac{|S_{R^*}(tc_i)|}{|U|}$, $i = 1, 2, \ldots, |U|$,

$$p\left(d_j \mid S_{R^*}(tc_i)\right) = \frac{p(S_{R^*}(tc_i), d_j)}{p(S_{R_*}(tc_i))} = \frac{|S_{R^*}(tc_i) \cap d_j|}{|S_{R^*}(tc_i)|},$$

$$i = 1, 2, \ldots, |U| \qquad j = 1, 2, \ldots, k$$

Example 2. Given a complete TCR in Table 1, we have

$$R^* = \{r_1, r_2, r_3, r_4, r_5, r_6, r_7, r_8, r_9, r_{10}, r_{11}, r_{12}\},$$

$$\frac{U}{IND(d)} = \{\{tc_2, tc_3, tc_4, tc_6, tc_7, tc_8, tc_9, tc_{10}, tc_{11}, tc_{12}\}, \{tc_1, tc_6\}\}$$

$$S_{R^*}(tc_1) = \{tc_1, tc_2, tc_3, tc_4, tc_5, tc_6, tc_{12}\} = S_{R^*}(tc_2) = S_{R^*}(tc_3) = S_{R^*}(tc_4)$$

$$= S_{R^*}(tc_5) = S_{R^*}(tc_6) = S_{R^*}(tc_{12}), S_{R^*}(tc_7) = \{tc_7\}, S_{R^*}(tc_8) = \{tc_8\},$$

$$S_{R^*}(tc_9) = \{tc_9\}, S_{R^*}(tc_{10}) = \{tc_{10}\}, S_{R^*}(tc_{11}) = \{tc_{11}\}$$

Definition 4. Let TCR, $TCR = (U, R^* \cup \{d\})$ be a complete Test Case Requirement and $B \subseteq R^*$. The requirement attributes set B is a relative reduction of R^* to decision attributes d, if and only if [6]:

(a) $EN(d|R^*) = EN(d|B)$;
(b) $B^{'} \subset B, EN(d|B^{'}) \neq EN(d|B)$.

$$EN(d \mid R^*) = -\left[7 * \frac{7}{12} \left(\frac{5}{7} \log_2 \frac{5}{7} \right) + \frac{2}{7} \log_2 \frac{2}{7} + \frac{1}{7}(1 \log_2 1) + \frac{1}{7}(1 \log_2 1) \right.$$

$$\left. + \frac{1}{7}(1 \log_2 1) + \frac{1}{7}(1 \log_2 1) + \frac{1}{7}(1 \log_2 1) \right]$$

$$= 5.0305$$

Example 3. Consequently, we can deduce other conditional entropy values with respect to the different condition requirements in the same way as follows:

$$R^* = \{r_1\}$$

$$S_{R^*}(tc_1) = \{tc_1, tc_2, tc_3, tc_4, tc_5, tc_6, tc_7, tc_8, tc_9, tc_{10}, Ttc_{11}tc_{12}\} = S_{R^*}(tc_2)$$

$$= S_{R^*}(tc_3) = S_{R^*}(tc_4) = S_{R^*}(tc_5) = S_{R^*}(tc_6) = S_{R^*}(tc_7) = S_{R^*}(tc_8) = S_{R^*}(tc_9) = S_{R^*}(tc_{10})$$

$$= S_{R^*}(tc_{11}) = S_{R^*}(tc_{12})$$

$$EN(d \mid R^*) = -\left[12 * \frac{12}{12} \left(\frac{10}{12} \log_2 \frac{10}{12} \right) + \frac{2}{12} \log_2 \frac{2}{12} \right]$$

$$= 7.8002$$

Example 4. The conditional entropy with respect to different combinations of six conditions attributes.

$$R = \{r_1, r_2, r_3, r_4, r_5, r_6\}$$

$$S_{R^*}(tc_1) = \{tc_1, tc_2, tc_3, tc_4, tc_5, tc_6, tc_8, tc_9, tc_{11}tc_{12}\} = S_{R^*}(tc_2)$$

$$= S_{R^*}(tc_3) = S_C(tc_4) = S_{R^*}(tc_5) = S_{R^*}(tc_6) = S_{R^*}(tc_8) = S_{R^*}(tc_9) = S_{R^*}(tc_{11}) = S_{R^*}(tc_{12}),$$

$$S_{R^*}(tc_7) = \{tc_7\}, S_{R^*}(tc_{10}) = \{tc_{10}\}$$

$$EN(d \mid C) = -\left[10 * \frac{10}{12}\left(\frac{8}{10}\log_2\frac{8}{10}\right) + \frac{2}{2}\log_2\frac{2}{2} + \frac{1}{1}\log_2\frac{1}{1} + 0 + \frac{1}{1}\log_2\frac{1}{1} + 0\right]$$
$$= 2.1462$$

Example 5. The conditional entropy with respect to different combinations of eight conditions attributes.

$$R* = \{r_2, r_5, r_6, r_7, r_8, r_9, r_{10}, r_{11}\}$$

$$S_{R*}(tc_1) = \{tc_1, tc_2, tc_3, tc_4, tc_5, tc_6, tc_{12}\} = S_{R*}(tc_2) = S_{R*}(tc_3) = S_{R*}(tc_4),$$

$$S_{R*}(tc_5) = S_{R*}(tc_6) = S_{R*}(tc_{12}), S_{R*}(tc_7) = \{tc_7\}, S_{R*}(tc_8) = \{tc_8\},$$

$$S_{R*}(tc_9) = \{tc_9\}, S_{R*}(tc_{10}) = \{tc_{10}\}, S_{R*}(tc_{11}) = \{tc_{11}\}$$

$$EN(d \mid C) = -\left[7 * \frac{7}{12}\left(\frac{5}{7}\log_2\frac{5}{7}\right) + \frac{2}{7}\log_2\frac{2}{7} + \frac{1}{7}(1\log_2 1) + \frac{1}{7}(1\log_2 1) + \frac{1}{7}(1\log_2 1)\right.$$
$$\left. + \frac{1}{7}(1\log_2 1) + \frac{1}{7}(1\log_2 1)\right] = 5.0305$$

Other conditional entropy for every combination of conditional requirements was calculated in the same way. Since, $EN(d \mid \{r_2, r_5, r_6, r_7, r_8, r_9, r_{10}, r_{11}\} = EN(d \mid R^*) = 5.035)$, the desired attribute set of selected $\{r_2, r_5, r_6, r_7, r_8, r_9, r_{10}, r_{11}\}$ was obtained, which was a relative reduction of the whole requirements set in R.

Hence, the final result of requirements selection was $\{r_2, r_5, r_6, r_7, r_8, r_9, r_{10}, r_{11}\}$. The Test Case classification and requirements selection produced a new Test Case Requirement, TCR' as in Table 2, which was smaller in size, compared with the actual TCR.

Table 2. A new test case requirement matrix table, TCR'

	r_2	r_5	r_6	r_7	r_8	r_9	r_{10}	r_{11}	Decision
tc_1, tc_5	1	1	1	0	0	0	0	0	Fail
tc_2, tc_3, tc_4, tc_6, tc_{12}	1	1	1	0	0	0	0	0	Pass
tc_7	0	0	0	1	0	0	0	0	Pass
tc_8	1	1	1	0	1	0	0	0	Pass
tc_9	1	1	1	0	0	1	0	0	Pass
tc_{10}	0	1	0	0	0	0	1	0	Pass
tc_{11}	1	1	1	0	0	0	0	1	Pass

Definition 5. Let a data set with n test cases and m requirements attributes. The size of the data was calculated as, $\text{SDataSet}(n, m) = n * m$.

Example: From Table 1, the $S_{\text{DataSet}}(12, 12) = 144$

4 Evaluation/Experimental Results

This section presented an evaluation results, showing the efficiency of the technique. The evaluation of the technique was executed on PC with 2.8 GHz CPU, 4.0 GB RAM and Windows 7 Professional. Data set used in this paper is from the Test Log of Automatic Teller Machine System Independent Verification and Validation Project (ATMS IV & V) by Malaysian Software Testing Board (*MSTB Lecturer Aid (Student)*). The evaluation included the size reduction and the execution time. In evaluating the effectiveness of the reduced size of test suite and requirements attribute, the reduction rate of proposed technique was calculated as described in Definitions 5 and 6:

Definition 6. Let $m(U)$ be the number of reduced test cases, $n(U)$ be the number of actual test cases. The reduction rate was calculated as follows:

$$\text{Reduction rate} = \frac{(n(U) - m(U))}{n(U)} * 100$$

Definition 7. Let $m(R)$ be a number of requirements reduction, $n(R)$ is the number of actual requirements in ATMS dataset $= n(R) = 12$. The reduction rate of requirements attribute can be calculated as:

$$\text{Reduction rate} = \frac{(n(R) - m(R))}{n(R)} * 100$$

Since time that proportional to the size of TCR, the processing time reduction rate is calculated as described in Definition 5. An evaluation was done, using the following data sets:

Table 3 shows that, Harris and Raju [10] reduced the highest number of test cases, compared with the other researchers (reduced from 10 to 3). However, the number of requirement attributes was not reduced, thus resulting in high execution time. Also, the fault detection capability was not verified. As a comparison, the proposed approach reduced test cases to 6 and reduced requirement attributes, also to 6, with 100% fault detection capability which resulted in lower execution time, making it better than the other techniques.

Table 3. Dataset 1 (10 test cases)

Authors	Year	Test suite minimization techniques	Comparative analysis							
			Number of actual test cases	Number of actual requirements	$S_{DataSet}$	Fault detection	Number of test case after reduced	Number of requirements after reduced	$S_{DataSet}$ after reduction	Fault detection capability (%)
Xu et al.	2012	Weighted Greedy Algorithm	10	10	100	All Pass	7	10	40	NA
Campos et al.	2013	RZOLTAR	10	10	100	All Pass	4	10	40	NA
Gotlieb et al.	2014	FLOWER	10	10	100	All Pass	8–9	10	80–90	100
Harris and Raju	2015	MFTS Algorithm	10	10	100	All Pass	3	10	30	NA
NFM et al.	2017	Test Case Requirement Reduction using Rough Set and Conditional Entropy	10	10	100	All Pass	6	6	36	100

5   Conclusions

Similarity relation was used to reduce the test cases, while conditional entropy was used to find minimum subset of requirements. The test cases and requirements in TCR table were reduced without affecting the fault detection capability of the system, which determined the results of the test cases either pass or fail. The results obtained showed that, the reduced test cases with the minimum selected requirements produced a greater reduction rate and reduced execution time, compared with the actual data. Thus, the new small-size TCR was produced after the implementation of the proposed technique, which contributed to shorter execution time. The concept of test case reduction was explored in this study to tackle the issue on finding the minimal hitting subset of test cases. In this research, the minimal subset of test cases and requirements was generated, using the concept of rough set and conditional entropy, with small data sets. Therefore, for large data sets, to reduce the time consumption of the testing process, in terms of classification and attribute selection, more applicable approaches, such as, heuristic algorithm are desirable for large-scale test cases. This issue needs to be investigated in the future. The scope of this study was to reduce the test case-requirements in TCR table. Although the reduction has not decreased the fault detection capability of the test cases, other quality metrics can also be used to validate the results and to ensure that the reduction has not affected the efficiency of the system.

Acknowledgment. We would like to acknowledge the support from Ministry of Higher Education in undertaking the research under Fundamental Research Grant Scheme (FRGS) vote number 1643 and 1610.

References

1. Singh, N.P., Mishra, R., Yadav, R.R.: Analytical review of test redundancy detection techniques. Int. J. Comput. Appl., 0975–8887 (2011)
2. Zeng, B., Tan, L.: Test criteria for model-checking-assisted test case generation: a computational study. Paper presented at 2012 IEEE 13th International Conference on Information Reuse and Integration (IRI) (2012)
3. Maung, H.M., Win, K.T.: Entropy based test cases reduction algorithm for user session based testing. Paper presented at the International Conference on Genetic and Evolutionary Computing (2015)
4. Mahapatra, R., Singh, J.: Improving the effectiveness of software testing through test case reduction. World Acad. Sci. Eng. Technol. 37, 345–350 (2008)
5. Deris, M.M., Abdullah, Z., Mamat, R., Yuan, Y.: A new limited tolerance relation for attribute selection in incomplete information systems. Paper presented at the 2015 12th International Conference on Fuzzy Systems and Knowledge Discovery (FSKD) (2015)
6. Yan, T., Han, C.Z.: A novel approach based on rough conditional entropy for attribute reduction. Paper presented at the Applied Mechanics and Materials (2014)
7. Xu, S., Miao, H., Gao, H.: Test suite reduction using weighted set covering techniques. Paper presented at 2012 13th ACIS International Conference on Software Engineering, Artificial Intelligence, Networking and Parallel & Distributed Computing (SNPD) (2012)
8. Campos, J., Abreu, R.: Leveraging a constraint solver for minimizing test suites. Paper presented at the 2013 13th International Conference on Quality Software (2013)
9. Gotlieb, A., Marijan, D.: FLOWER: optimal test suite reduction as a network maximum flow. Paper presented at the Proceedings of the 2014 International Symposium on Software Testing and Analysis (2014)
10. Harris, P., Raju, N.: Towards test suite reduction using maximal frequent data mining concept. Int. J. Comput. Appl. Technol. 52(1), 48–58 (2015)
11. Harrold, M.J., Gupta, R., Soffa, M.L.: A methodology for controlling the size of a test suite. ACM Trans. Softw. Eng. Methodol. (TOSEM) 2(3), 270–285 (1993)
12. Parsa, S., Khalilian, A.: On the optimization approach towards test suite minimization. Int. J. Softw. Eng. Appl. 4(1), 15–28 (2010)

A Framework to Detect Compromised Websites Using Link Structure Anomalies

Patchmuthu Ravi Kumar[1(✉)], Perianayagam Herbert Raj[2],
and Perianayagam Jelciana[3]

[1] School of Computing and Informatics,
Universiti Teknologi Brunei, Gadong, Brunei
ravi.patchmuthu@utb.edu.bn
[2] School of Information and Communication Technology,
IBTE, SB Campus, Gadong, Brunei
herbert.raj@ibte.edu.bn
[3] Gadong, Brunei
jelcianap@gmail.com

Abstract. Compromised or malicious websites are a serious threat to cyber security. Malicious users prefer to do malicious activities like phishing, spamming, etc., using compromised websites because they can mask their original identities behind these compromised sites. Compromised websites are more difficult to detect than malicious websites because compromised websites work in masquerade mode. This is one of the main reasons for us to take this topic as our research. This paper introduces the related work first and then introduces a framework to detect compromised websites using link structure analysis. One of the most popular link structures based ranking algorithm used by the Google search engine algorithm called PageRank, is implemented in our experiment using the Java language, computation is done before a website is compromised and after a website is compromised and the results are compared. The results show that when a website is compromised, its PageRank can go do down to indicate that this website is compromised.

Keywords: Compromised websites · Link structure · Malicious websites
Spamming · Redirection · Cyber attack · Web graph · PageRank

1 Introduction

Internet and World Wide Web (WWW) is becoming part and parcel of our day-to-day life. Spammers and hackers take advantage of the colossal growth and the human dependency of WWW to do malicious activities in the Web. According to WorldWideWebSize.com, there are 4.47 billion pages [1] as of 28th June 2018. Due to the amazing growth of WWW and internet, many of our day-to-day activities like shopping, buying tickets, banking, education, paying bills and taxes, etc., are shifted to

P. Jelciana—IT Consultant.

© Springer Nature Switzerland AG 2019
S. Omar et al. (Eds.): CIIS 2018, AISC 888, pp. 72–84, 2019.
https://doi.org/10.1007/978-3-030-03302-6_7

the Web for convenience and ease. As more and more commercial and business activities are moved to the web, the website owners are more concerned with their rankings in the search engine results. Benign websites use legitimate methods to improve their ranking while malicious websites use illegitimate methods to improve their rankings. Web is always under a constant threat from malicious actors because of the amount of information that has accumulated in the Web. The threat could be a trivial one like e-mail spam or it could be a more serious one like theft of money or identity and/or loss of a business to a competitor. Malicious Websites (URLs) are the base platforms for all these counterfeit activities such as spamming, phishing and malware injection. There are a wide variety of attacks that can happen in the cyber space such as hacking, social engineering attacks, drive-by exploits, SQL injections, denial-of-service, distributed denial-of-service, man-in-the middle attack and many more [2]. The software which is dedicated for doing all these activities is called bad ware [3] and more popularly known as malware.

Malicious websites are created intentionally for doing illegal activities in the web and the compromised websites are created as benign sites, but they are being compromised by spammers/hackers for the purpose of doing cyber-attacks. Hackers prefer compromised websites because they can hide their real identity and do all the counterfeit activities. Also, the hackers get free hosting, bandwidth and computing power. Generally, the compromised websites are difficult to identify with most of the web security software and these sites are less likely to be blocked by them. This research is based on studying the link structure properties like in-degree and out-degree of the compromised websites and compute the rank of the websites before and after it is being compromised.

This paper organized as follows. Section 2 provides the background and reviews the related work. Section 3 provides a framework to detect the compromise websites using link structure anomalies. In Sect. 4, a simple experiment is done to extract in-degree, out-degree and PageRank is computed before and after a website is compromised. Section 5 concludes this paper with a couple of future directions.

2   Background and Related Work

There are a number of tools available online to detect malicious websites like McAfee Site Advisor, Norton Safe Web, Sucuri Site Check, URL Blacklist, VirusTotal, Google Malware Checker etc. A number of researches has been done or is happening to find out the malicious websites using different methodologies [4–9] but only a couple of researches [10–12] have been done to detect the compromised websites. Canali et al. [10] created vulnerable websites, hosted on shared web hosting and they ran five different attacks against them. According to their result analysis, most of the shared hosting providers are unable to detect even the simplest form of malicious activity. Soska and Christin [11] used classification method to find vulnerable websites before they turn into malicious. Shibahara et al. [12] used classifiers to detect not only malicious websites but also compromised websites. The following sub section describes the characteristics of compromised websites, reasons for compromise and the methods used for compromise.

2.1   Compromised Websites

A compromised website is a website, whose Confidentiality, Integrity and Availability (CIA) has been seriously impacted by untrusted sources either intentionally or unintentionally. Malicious actors can compromise a website either manually or through automation. Commtouch.com [3] did a survey on owner's perspective of the compromised websites using many useful parameters. The following are the reasons for a website is being compromised:

- Universal Resource Locator (URL) Redirect – one of the basic reasons for compromising a website is to redirect it to a master URL by hiding a HTML code in the compromised website and forcing the site as 'front door' to do all malicious activities. The master URL has spam or malware. It is normally done through email and asks the user to click the link and then redirects to a malicious site.
- Hosting Malware – malicious actors employ hidden scripts in themes especially in WordPress or ask an email recipient to download a malware file which is hosted on a compromised website.
- Hosting Phishing, Spam pages and illicit materials – malicious actors use compromised websites (static pages) to advertise spam products (replicas, pharmaceuticals, other boosters, etc.) that act as phishing pages for companies. Phishing is mainly used for stealing personal information like login credentials etc. for cybercrime purposes.
- Vandalism – malicious actors try to embrace the site owners for political reasons, business competition purposes etc. It is generally called as hacktivism.
- Other Activities – like drive-by downloads, other scripting activities and any other methods come under this.

2.2   Symptoms of a Compromised Website

There are many symptoms of a compromised website. The following are some of the important signs [13] which will be useful for the site owners to act and bring the compromised site back to a normal site:

- Browser Warning – browser warnings are the first notification when a user visits a compromised or hacked site. It is better to avoid visiting these sites. According to Google, the False Positive (FP) rate for this warning is extremely low (meaning is that the site is compromised or hacked). The following Fig. 1 shows a warning sign from Google.
- Spam Mail from your Website – automatic spam or unwanted email or text that goes to other users' devices from your domain email indicates that your site is compromised.
- Slow or Unresponsive Site – takes more time to load or even doesn't respond and produces HTTP server 500/503 errors as shown in Fig. 2. It could be a possible compromise.
- Redirection – many types of compromise can happen to a site using redirection. The common one is when a user types a domain in the URL, the site opens temporarily first and then redirects to some strange looking sites selling all sorts of

Fig. 1. Google site warning

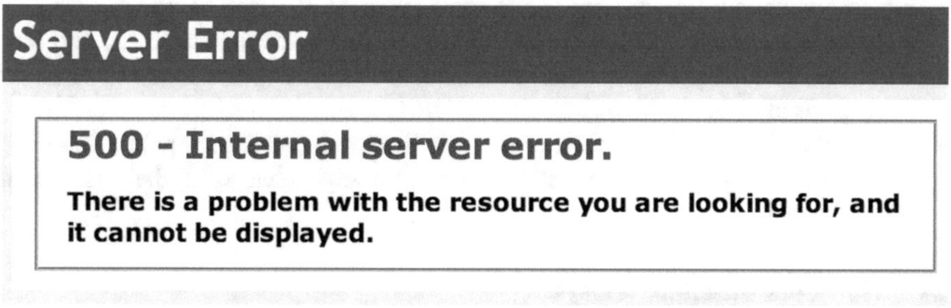

Server Error

500 - Internal server error.

There is a problem with the resource you are looking for, and it cannot be displayed.

Fig. 2. Internal server error (possible compromise)

counterfeit/dodgy products or porn sites. This is a confirmed sign that your site is compromised. The second one is using spam email's link and the third one is redirected internet searches where the hackers can redirect the Search Engine Result Pages (SERP) when a user clicks on a particular result (hackers get paid for the user clicks). Fourth one is malwares that can add bogus toolbar programs in the browser settings to redirect into different sites. Fifth one is Pharma Hack or Search Engine Optimization (SEO) spam where hackers exploit the vulnerable code to distribute counterfeit/cheap drugs or pharmaceutical contents.

- Frequent Random Popups – frequent random popups from the browser is also a sign that your website is being compromised. It occurs due to the bogus toolbars.
- Drive-by Download – when users simply view an infected site, the malicious actors can compromise a website by injecting malicious objects like JavaScript code to iFrames etc. and downloads malware into the client's computer without the knowledge of the user.
- Increase in the Traffic – a sudden increase in the popularity of a website shows that the website is compromised. This could be due to redirection by malicious actors. Studying the external outbound links of a website can disclose this kind of redirection. This is one of the main studies of this paper.

2.3   How Are Websites Compromised

Malicious actors always use new flaws, exploits and social engineering tricks to compromise a website. According to [3], 63% of the website owners don't know how their websites are compromised. There are many ways a website can be compromised. The following are some of the popular methods [14] used by malicious actors:

- Malware Exploit Kits – malware stands for malicious software, which is specifically designed to disturb, damage or gain access to a computer system. There are many types and variations of malware. To compromise a website, malwares like drive-by download, exploit kits, JavaScript infections, malvertising (malicious advertising), URL injections, malicious redirects, browser hijackers etc. can be used.
- Insecure/Out of date Software – malicious actors exploit out of date or insecure software like operating system, browser and other software.
- Stolen credentials – malicious actors use malicious activities to retrieve the user credentials (username/password) and then use these credentials to enter into their sites and then they can do any activities. In other words, those websites are compromised.
- Public Wi-Fi and PCs – when users use their site using any public Wi-Fi or computers without much protection, there is a high chance of their site being compromised.

2.4   Web Link Structure Analysis

Websites are organized as a set of pages and connected through hyperlinks. Researchers used this basic web model and represented the website as directed graphs [15–17]. Let $G = (V, E)$ where a vertex or a node or a page $p \in V (G)$ represents a webpage and an edge $e = \{i, j\} \in E(G)$, represents a hyperlink from page i to page j. This model is more popularly called a web graph and it is used by researchers and Search Engine Optimization (SEO) experts to improve the quality of search engine results and the rank of a website. Malicious actors also use this link structure to improve their website ranking and redirect the users into malicious websites. The main focus of this paper is to find out the compromised websites using link structure anomalies. A website has two types of links: incoming links and outgoing links. Incoming links are the hyperlinks that are coming into a page and outgoing links are the hyperlinks that are going out from a page. Again, these incoming and outgoing links can be internal links and external links. The internal links are the hyperlinks within a website connecting to or from one page to another page. External links are the hyperlinks which are connecting the website (page) to other domains. We are interested only in the external links (both incoming and outgoing links). For this research, we used a couple of definitions related to directed graph which is shown in the next section.

3   Framework to Detect Link Structure Anomalies

The following are the definitions related to our proposed methodology. These definitions are used in the PageRank algorithm which is used in the experiment later in Sect. 4.

Definition 1: The in-degree (*id*) of a page *i* is nothing but the number of incoming links as shown in Eq. (1).

$$id(i) = \sum_i E_{ji} \tag{1}$$

Definition 2: The out-degree (*od*) of a page *i* is nothing but the number of outgoing links as shown in Eq. (2).

$$od(i) = \sum_i E_{ij} \tag{2}$$

Definition 3: The adjacency matrix A_{ij} is nothing but a connectivity matrix or link matrix which can be created using the Eq. (3) as shown below.

$$A_{i,j} = \begin{cases} 1 & \text{if } (i,j) \in E \\ 0 & \text{otherwise} \end{cases} \tag{3}$$

The generalized $n \times n$ adjacency matrix A (row matrix) for a directed graph is shown below:

$$A = \begin{bmatrix} a_{1,1} & a_{1,2} & \cdots & a_{1,n} \\ a_{2,1} & a_{2,2} & \cdots & a_{2,n} \\ \vdots & \vdots & \ddots & \vdots \\ a_{n,1} & a_{n,2} & \cdots & a_{n,n} \end{bmatrix}$$

Definition 4: A *transition probability matrix* (*P*) can be defined as $P_{ij} = E_{ij}/od(i)$ when deg(i) > 0 [18]. For a stochastic matrix, the *i*th row sum to 1. Google uses this stochastic matrix while calculating the PageRank of a web page [19]. Let *P* be an $n \times n$ matrix whose element P_{ij} is the probability of moving from page *i* to page *j* in one step. The transition probability matrix can be created using the Eq. (4) as shown below:

$$P_{i,j} = \begin{cases} a_{i,j}/s_j & \text{if } s_j \neq 0 \\ 0 & \text{if } s_j = 0 \end{cases} \tag{4}$$

In the above Eq. (4), $a_{i,j}$ is the connection from page *j* to page *i* and if there is a connection then $a_{i,j} = 1$, otherwise $a_{i,j} = 0$. s_j is the row sum of the pages and $s_j =$ and it is nothing but the out-degree of page *j*. The generalized transition probability matrix for a directed graph is shown below:

$$P = \begin{bmatrix} p_{1,1} & p_{1,2} & \cdots & p_{1,n} \\ p_{2,1} & p_{2,2} & \cdots & p_{2,n} \\ \vdots & \vdots & \ddots & \vdots \\ p_{n,1} & p_{n,2} & \cdots & p_{n,n} \end{bmatrix}$$

Definition 5: We used the popular PageRank algorithm (used by the Google Search engine algorithm to rank web pages) to compute the rank of our sample web sites. PageRank formula is shown below in Eq. (5).

$$PR(i) = d \sum_{j \in id_i} \frac{PR_j}{od_j} + (1 - d) \tag{5}$$

In the above PageRank formula, PageRank (*PR*) of a page *i* can be computed by taking all the incoming pages to page *i*, i.e. id_i. Here, *d* is a damping factor which is normally between $0 < d < 1$ and od_j is the number of outgoing links of page *j*.

In-degree (*id*) is used to monitor the traffic of a website. It is using the external incoming links. This can tell the sudden popularity of a website. Generally, search engines use the external incoming links (back links) as one of the important factors for ranking a website. Out-degree (*od*) is used to find out the number of links a website is connecting to other pages (external) and this will be useful in finding the redirection.

Our proposed framework to detect the compromised websites using link structure analysis is shown below in Fig. 3.

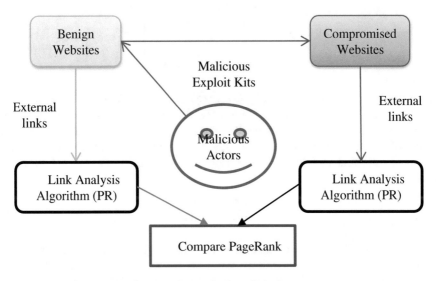

Fig. 3. Proposed framework to detect link anomalies

Malicious actors, using the exploit kits compromise benign websites. Our link analysis algorithm (PageRank) extracts both *id*s and *od*s (external) before a website is compromised and after a website is compromised, compute PageRank and compares. If there is a sudden increase in the number of incoming links, then there is a possibility that the site is being compromised or something fishy is going on in that site. Other parameters like URLs, host-based features (HTTP, DNS, IP address etc.), web contents, visual features of a site and other parameters [2, 11, 20] are used by researchers to find out malicious web sites.

3.1   Example on Link Structure Analysis

Let us a take a sample web graph (WG1) with 4 websites and 9 edges as shown in Fig. 4. Assume these four websites are benign websites and are related to one common topic, for example, Internet of Things (IoT). Let us take that the website 2 is compromised by malicious actors and redirected to pharmaceutical marketing websites 5 and 6 as shown in Fig. 5. Our experiments compute the *id*, *od*, adjacency matrix (*A*), transition probability matrix (*P*) and PageRank of those four websites (benign sites) and also for the 6 websites (after website 2 is compromised) and compare the results in the experiments section.

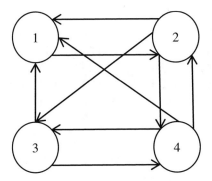

Fig. 4. A sample web graph (WG1) for related group of IoT

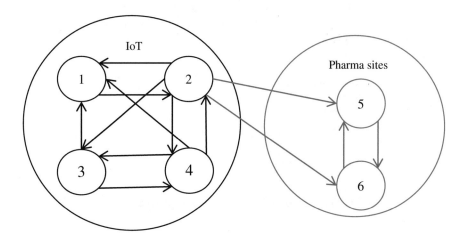

Fig. 5. Sample web graph (WG2) after compromise (website 2)

The adjacency matrix (*A1*) is computed and shown below by applying the Eq. (3) into the sample web graph (WG1) of Fig. 4. The adjacency matrix (*A2*) is computed and shown below by applying the Eq. (3) into the sample web graph (WG2) of Fig. 5.

$$A1 = \begin{bmatrix} 0 & 1 & 0 & 0 \\ 1 & 0 & 1 & 1 \\ 1 & 0 & 0 & 1 \\ 1 & 1 & 1 & 0 \end{bmatrix} \quad A2 = \begin{bmatrix} 0 & 1 & 0 & 0 & 0 & 0 \\ 1 & 0 & 1 & 1 & 1 & 1 \\ 1 & 0 & 0 & 1 & 0 & 0 \\ 1 & 1 & 1 & 0 & 0 & 0 \\ 0 & 0 & 0 & 0 & 0 & 1 \\ 0 & 0 & 0 & 0 & 1 & 0 \end{bmatrix}$$

The transition probability matrix ($P1$) is computed and shown below by applying the Eq. (4) into the sample web graph (WG1) of Fig. 4. The transition probability matrix ($P2$) is computed and shown below by applying the Eq. (4) into the sample web graph (WG2) of Fig. 5.

$$P1 = \begin{bmatrix} 0 & 1 & 0 & 0 \\ 1/3 & 0 & 1/3 & 1/3 \\ 1/2 & 0 & 0 & 1/2 \\ 1/3 & 1/3 & 1/3 & 0 \end{bmatrix} \quad P2 = \begin{bmatrix} 0 & 1 & 0 & 0 & 0 & 0 \\ 2/5 & 0 & 2/5 & 2/5 & 2/5 & 2/5 \\ 1/2 & 0 & 0 & 1/2 & 0 & 0 \\ 1/3 & 1/3 & 1/3 & 0 & 0 & 0 \\ 0 & 0 & 0 & 0 & 0 & 1 \\ 0 & 0 & 0 & 0 & 1 & 0 \end{bmatrix}$$

4   Experiments

We developed a small PageRank program using Java to calculate the PageRank of the sample websites shown in Fig. 4. The input to the program is the adjacency matrix and the program will calculate the PageRank of the websites. The print screen for the input to the program is shown in Fig. 6.

Fig. 6. Input to the PageRank program (4 Benign websites)

The following Fig. 7 shows the computed PageRank for the 4 websites. The ranks of the websites are shown in Table 1.

Figures 8 and 9 shows the input and the PageRank output after the website 2 is compromised and redirected to websites 5 and 6. The following Table 1 shows the in-degree (*id*), the out-degree (*od*) and PageRank of the sample websites (4 sites) before compromise as shown in Fig. 4 and after compromise as shown in Fig. 5. But in a real web scenario, there are hundreds and thousands of in-degrees and out-degrees

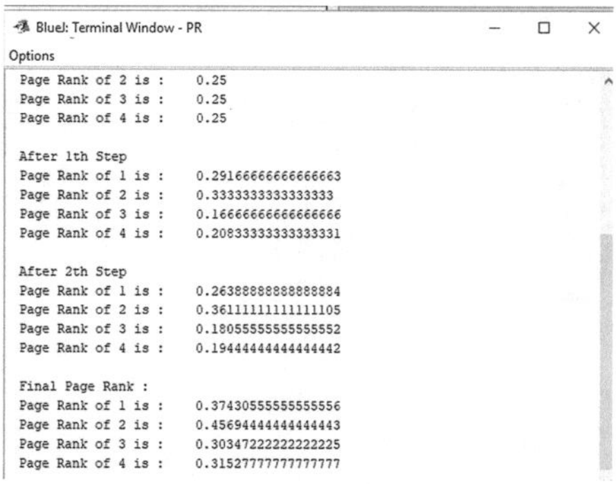

```
Blue): Terminal Window - PR                              —  □  ×
Options
   Page Rank of 2 is :    0.25
   Page Rank of 3 is :    0.25
   Page Rank of 4 is :    0.25

   After 1th Step
   Page Rank of 1 is :    0.2916666666666663
   Page Rank of 2 is :    0.3333333333333333
   Page Rank of 3 is :    0.1666666666666666
   Page Rank of 4 is :    0.2083333333333331

   After 2th Step
   Page Rank of 1 is :    0.2638888888888884
   Page Rank of 2 is :    0.3611111111111105
   Page Rank of 3 is :    0.1805555555555552
   Page Rank of 4 is :    0.1944444444444442

   Final Page Rank :
   Page Rank of 1 is :    0.3743055555555556
   Page Rank of 2 is :    0.4569444444444443
   Page Rank of 3 is :    0.3034722222222225
   Page Rank of 4 is :    0.3152777777777777
```

Fig. 7. Computed PageRank for the four benign websites

Table 1. In-degree, out-degree and PageRank before and after compromise

Website	Before compromise			After compromise		
	Id	Od	PageRank	Id	Od	PageRank
1	3	1	0.37	3	1	0.26
2	2	3	0.46	2	**5**	**0.33**
3	2	2	0.30	2	2	0.22
4	2	3	0.32	2	3	0.23
5	–	–	–	2	1	0.19
6	–	–	–	2	1	0.22

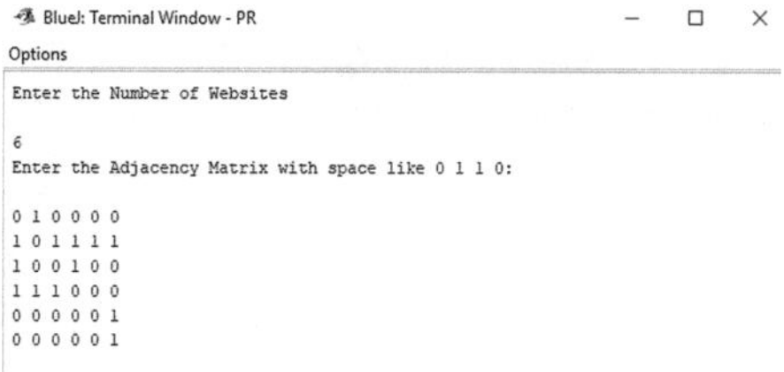

```
Blue): Terminal Window - PR                              —  □  ×
Options
Enter the Number of Websites

6
Enter the Adjacency Matrix with space like 0 1 1 0:

0 1 0 0 0 0
1 0 1 1 1 1
1 0 0 1 0 0
1 1 1 0 0 0
0 0 0 0 0 1
0 0 0 0 0 1
```

Fig. 8. Input to the PageRank program (compromised website)

depending on how popular a website is. Also computing the PageRank is more complex and time consuming.

The above Table 1 and Fig. 10 shows that when a website is compromised for redirection, its PageRank goes down. Here website 2 is compromised and redirected to pharmaceutical sites and its PageRank goes down from 0.46 to 0.33. The ranks of other websites which are connected to the compromised websites also get affected. Also, the out-degree (*od*) count increases suddenly. This *od* and PageRank can give an indication to the web administrators that their sites are under compromise and they can do further

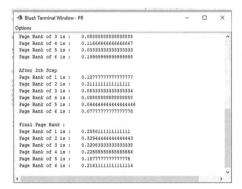

Fig. 9. Computed PageRank for the six websites after compromise

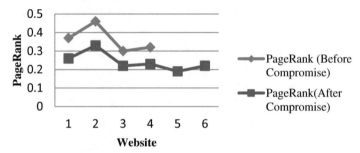

Fig. 10. Comparison of PageRank before and after compromise

analysis with VirusTotal.com and other proven methods for further investigation. Anchor text of the external incoming and outgoing links can provide more information on the type of links.

5   Conclusion

This paper takes a new initiative to detect compromised websites by analyzing the link structure of the websites and computing the PageRank. Compromised websites are not easy to detect because they always operate in a masquerade mode. They are used by malicious actors for redirection to adware sites, to deliver malicious content, to host phishing pages to steal personal information etc. There are many security tools available to detect malicious websites and also a number of researches have been done on it. We did a small experiment using PageRank algorithm to detect the link structure anomalies which in turn can alert the website owners that their websites are in the process of being compromised. Due to time and resource constraints, live experiments cannot be done on the internet. This can be done in the future and other machine learning algorithms can be applied to automate this process.

References

1. WorldWideWebSize.com: The size of the world wide web (the internet). http://www.worldwidewebsize.com/. Accessed 29 June 2018
2. Sahoo, D., Liu, C., Hoi, S.C.H.: Malicious URL detection using machine learning: a survey. arXiv:1701.07179v2 [cs.LG] (2017)
3. Stopbadware.org: Compromised websites an owner's perspective. https://www.stopbadware.org/files/compromised-websites-an-owners-perspective.pdf. Accessed 14 June 2018
4. Choi, H., Zhu, B.B., Lee, H.: Detecting malicious web links and identifying their attack types. In: 2nd USENIX Conference on Web Application Development, USA, 15–16 June 2011
5. Chiba, D., Tobe, K., Mori, T., Goto, S.: Detecting malicious websites by learning IP address features. In: 12th International Symposium on Applications and Internet (IEEE/IPSJ), Turkey, 16–20 July 2012
6. Justin, M., Lawrence, K.S., Stefan, S., Geoffrey, M.V.: Learning to detect malicious URLs. ACM Trans. Intell. Syst. Technol. (TIST) **30**(3), 30:1–30:23 (2011)
7. Ramesh, G., Krishnamurthi, I., Kumar, K.S.S.: An efficacious method for detecting phishing webpages through target domain identification. Elsevier Decis. Support Syst. **61**, 12–22 (2014)
8. Vanhoenshoven, F., Napoles, G., Falcon, R., Vanhoof, K., Koppen, M.: Detecting malicious URLs using machine learning techniques. In: IEEE Symposium Series on Computational Intelligence (SSCI 2016), Greece, 6–9 December 2016
9. Wang, Y., Cai, W.D., Wei, P.C.: Deep learning approach for detecting malicious Javascript code. Secur. Commun. Networks. Wiley Online Libr. **9**(11), 1520–1534 (2016). https://doi.org/10.1002/sec.1441
10. Canali, D., Balzarotti, D., Francillon, A.: The role of web hosting providers in detecting compromised websites. In: 22nd International Conference on World Wide Web Conference (WWW 2013), Brazil, 13–17 May 2013
11. Soska, K., Christin, N.: 23rd USENIX Security Symposium (USENIX Security 2014), USA, 20–22 August 2014
12. Shibahara, T., Takata, Y., Akiyama, M., Yagi, T., Yada, T.: Detecting malicious websites by integrating Malicious, Benign, and compromised redirection subgraph similarities. In: IEEE 41st Annual Computer Software and Applications Conference (COMPSAC), vol. 1, pp. 655–664 (2017)
13. McEntee, D.: The 10 signs you have a compromised website. https://www.webwatchdog.io/2016/10/21/the-10-signs-youve-a-hacked-or-compromised-website/. Accessed 27 June 2018
14. Cucu, P.: How malicious websites infect you in unexpected ways. https://heimdalsecurity.com/blog/malicious-websites/. Accessed 30 June 2018
15. Ravi Kumar, P., Ashutosh, K.S., Anand, M.: A new algorithm for detection of link spam contributed by zero-out-link pages. Turk. J. Electr. Eng. & Comput. Sci. (TUBITAK) **24**, 2106–2123 (2016). https://doi.org/10.3906/elk-1401-202
16. Ravi Kumar, P., Alex Goh, G.L., Ashutosh, K.S., Anand, M.: Efficient methodologies to determine the relevancy of hanging pages using stability analysis. Cybern. Syst. (2016). https://doi.org/10.1080/01969722.2016.1187030
17. Eduarda, M.R., Natasha, M.F., Martin, H., Gavin, S.: Link structure graph for representing and analyzing web sites. Microsoft Research. (MSR-TR-2006–94) (2006)

18. Ravi Kumar, P., Alex Goh, K.L., Ashutosh, K.S.: Application of Markov Chain in the PageRank algorithm. Pertanika J. Sci. Technol. **21**(2), 541–554 (2013)
19. Brin, S., Page, L.: The anatomy of a large scale hypertextual web search engine. Comput. Netw. ISDN Syst. **30**(1–7), 107–117 (1998)
20. Gianluca, S., Christopher, K., Giovanni, V.: ACM SIGSAC Conference on Computer & Communications Security, CCS 2013, Germany, 4–8 November 2013

Network Centric Computing

A Performance Study of High-End Fog and Fog Cluster in iFogSim

Fatin Hamadah Rahman[1](✉), Thien Wan Au[1], S. H. Shah Newaz[1,2], and Wida Susanty Haji Suhaili[1]

[1] School of Computing and Informatics, Universiti Teknologi Brunei,
Jalan Tungku Link, Gadong BE1410, Brunei Darussalam
p20171005@student.utb.edu.bn
[2] KAIST Institute for Information Technology Convergence,
291 Daehak-ro, Yuseong-gu, Daejeon 34141, South Korea
{twan.au,shah.newaz,wida.suhaili}@utb.edu.bn

Abstract. Fog computing is introduced to help leverage the processing burden in cloud as it can no longer sustain the ever-growing volume, velocity and variety of the IoT-generated data. Existing studies have shown that fog computing is able to reduce the latency and provide other benefits in the IoT-fog-cloud environment. While fog is heterogeneous since can be any device with computing, networking and storage ability, its scalability must be ensured. Currently, not many studies have been conducted to see the performance of fog in different scalability approaches i.e. scaling up and scaling down. This paper provides a brief explanation on iFogSim, which is a Java-based program that allows modelling and simulation of fog computing environments. The iFogSim is used in this study to simulate the fog environment in scaling up and scaling out approaches running in five configuration settings. In the scaling out approach, it presents a cluster of fogs with similar specifications and the scaling up approach presents a high-end fog with greater capabilities than the fogs in the first approach. Our initial findings delineate that the scaling out approach gives a better result in reducing the cost of execution in cloud. In this paper, we provide an insightful discussion on the strength and weakness in these two approaches. This would open up new avenues of further research.

Keywords: Fog computing · Scalability · iFogSim

1 Introduction

Executing domestic tasks are effortless nowadays from brewing coffee the moment a person wakes up to receiving real-time inventory updates. This is all made possible with the Internet of Things (IoT) that has positively affect people's lives globally. With the IoT growing every day, the Business Insider reported that by 2020, 34 billion devices will be connected to the Internet, 24 billion of which will be IoT devices [19]. Consequently, the need for data to be processed

© Springer Nature Switzerland AG 2019
S. Omar et al. (Eds.): CIIS 2018, AISC 888, pp. 87–96, 2019.
https://doi.org/10.1007/978-3-030-03302-6_8

quickly also increases. While cloud computing contributes tremendously to the success of IoT, however, it can no longer sustain in processing the ever-growing IoT demand whilst meeting the stringent latency requirement. As a solution, fog computing is introduced to help leverage the burden in cloud by allowing the processing to be done locally [5]. To illustrate how a network bandwidth reduction can be achieved by implementing fog computing, let us consider a city surveillance scenario. If the city employs 5 megapixel IP surveillance cameras, streaming it from the cloud using the T1 connection with transmission rate of 1.54 Mbps would consume a total of 3.9Tb data for one camera alone. If the city municipal were to deploy multiple advanced cameras, additional bandwidth would be required. Mitigating part of the tasks to fog not only reduces the burden in cloud, much of the processing can be done locally in the fog as well. From the various works such as in [10,18], it is obvious that implementing the fog layer in the IoT-cloud environment have brought a significant reduction in energy consumption, latency and network usage.

While the fog layer is heterogeneous in a way that a fog can be any device that has networking, computing and storage capabilities, ranging from servers, routers, set-top boxes to access points, the fog layer must also be able to dynamically scale depending on the network needs. Although there are studies that have looked into scalability in the fog [8,16,20], these studies did not specifically look into the scaling out (horizontal) and scaling up (vertical) approaches of the fog, and how these approaches would impact the cloud. Scaling up or scaling out has their own benefits and drawbacks. Thus, it is important to identify the appropriate scaling approach in order to maximize the overall performance without compromising other equally important factors. To date, there are no studies that have compared the scaling performance in the fog layer domain. Hence, the aim of this paper is to see the effect of implementing scaling up and scaling out approaches in fog and gauge the performance in terms of the cost of execution.

The structure of the paper is as follows: Sect. 2 describes the background study of the fog computing paradigm and scalability. Section 3 elaborates on iFogSim and the experiment configurations. Section 4 shows the results and discussions. Finally, conclusions and future works are presented in Sect. 5.

2   Related Work

2.1   Fog Computing Paradigm

It is indisputable that research interest in the fog computing domain has been getting attention, with possible appealing use cases. The OpenFog consortium has laid out several use cases that includes a fog-based drone delivery services [14] and smart cities [15]. Additionally, the existence of fog computing is applicable to a wide variety of applications such as smart grids, wireless sensor and actuator networks (WSAN), decentralized smart building control (DSBC), IoT and Cyber-physical systems (CPSs), and connected vehicles [4,17]. Currently, there are only a handful simulators capable of simulating a fog environment, one of which is iFogSim [7], a Java-based program that is an extension to

CloudSim. Various existing studies have used iFogSim to explore the capabilities of fog. The authors in [9] have extended the iFogSim to provide mobility support through migration of virtual machines between cloudlets. Along with their proposed migration policy, MyiFogSim can be used to analyze the policy impact on application quality of service. On the other hand, authors in [12] proposed a latency-aware application module management policy for fog environment that meets the diverse service delivery latency and amount of data signals to be processed in per unit time for different applications. An energy-aware allocation strategy for placing application modules (tasks) on fog devices is proposed by authors in [10]. Meanwhile, the authors in [18] presented a Module Mapping Algorithm for efficient utilization of resources in the network infrastructure by efficiently deploying Application Modules in Fog-Cloud Infrastructure for IoT based applications. Despite the studies conducted in fog, it is apparent that fog's scalability has not been widely explored.

Scalability is defined as the computer system, network or application ability to handle increasing amount of work, both in terms of processing power as well as storage resources [16]. A system is said to be scalable if the performance improves after additional resources are added and it can be scaled in two ways i.e. scaling up (vertically) or scaling out (horizontally) [1]. Scaling up adds resources to a single node in a system such as adding storage, processors or memory to a single computer while scaling out adds more nodes to a system, such as adding a new computer to a distributed software application. However, both approaches have their own advantages and disadvantages as elaborated in Table 1. Preference as to whether the use of scaling up or scaling out is well-suited for a system is context-dependent. For instance, scaling out approach is preferred by Google where they have demonstrated that 15,000 commodity-class PCs with fault-tolerant software is more cost-effective than a comparable system built out of a smaller number of high-end servers [3]. Data processing on the other hand, such that of Hadoop MapRduce are often better served by a scale-up server than a scale-out cluster [2].

Table 1. Advantages and disadvantages of scaling up and scaling out.

Scale up	**Advantages:**
	1. Less physical space and cooling is needed
	2. Easier maintenance
	Disadvantages:
	1. In some cases, upgrading some resources could be expensive
	2. Less availability and risk having a single point of failure
	Hence, increasing chances of server downtime and affecting the Service Level Agreement
Scale out	**Advantages:**
	1. Provides redundancy thus increasing the availability of the system
	Disadvantages:
	1. Troubleshooting and maintenance can be challenging and time-consuming
	2. It would require more space and cooling

3   Simulation

In this section, the scalability in fog is tested using iFogSim as it allows the flexibility of configuring the fog devices' specifications. To gauge the performance between the two approaches, the cloud execution cost metric is used.

3.1   IFogSim

The classes in iFogSim are annotated in a way that users without prior knowledge of CloudSim can easily define the infrastructure, service placement and resource allocation policies for Fog computing. The main java classes is depicted in Fig. 1 which comprise of FogDevice, Controller, ModuleMapping, ModulePlacementMapping, ModukePlacement, Application, AppEdge, AppLoop and AppModule. The Controller, ModuleMapping, ModulePlacementMapping and ModukePlacement belong to the same package. Meanwhile Application, AppEdge, AppLoop and AppModule belong to another package. In order to run the program in the iFogSim, another java class that is not shown in the diagram named Simulation is created. The details of the classes are further explained in Table 2 [7]. Nevertheless, iFogSim is packaged with two application module placement strategies; cloud-only placement and edge-ward placement as elaborated below [11]:

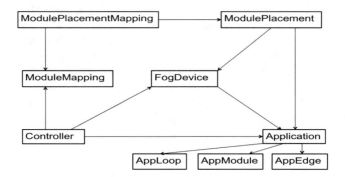

Fig. 1. Main iFogSim java classes and their relationships.

   A. Cloud-only placement: The cloud-only placement strategy is based on the traditional cloud- based implementation of applications where all modules of an application run in data centers. The sense-process-actuate loop in such applications are implemented by having sensors transmitting sensed data to the cloud where it is processed and actuators are informed if action is required.

   B. Edge-ward placement: Edge-ward placement strategy is inclined towards the deployment of application modules close to the edge of the network. However, devices close to the edge of the network may not be computationally powerful enough to host all operators of the application. In such a situation, the strategy iterates on fog devices towards cloud and tries to place remaining operators on alternative devices.

Table 2. iFogSim's java classes.

Java class	Remarks
FogDevice	The hardware characteristics of Fog devices and their connections to other Fog devices, sensors and actuators are specified in this class. The major attributes of the fog device include accessible memory, processor, storage size, uplink and downlink bandwidths. The methods in this class define how the resources of a fog device are scheduled between application modules running on it and how modules are deployed and decommissioned on them
Controller	The Controller object launches the AppModules on their assigned Fog devices following the placement information provided by Module Mapping object and periodically manages the resources of Fog devices
ModuleMapping	This class maps the node name to the module name
ModulePlacementMapping	Provides the placement mapping information
ModulePlacement	Contains the abstract placement policy that needs to be extended for integrating new policies
Application	Represents an application in the Distributed Dataflow Model
AppEdge	denotes the data-dependency between a pair of application modules and represents a directed edge in the application model
AppLoop	An additional class, used for specifying the process-control loops of interest to the user
AppModule	Represents the processing elements of fog applications
Simulation	This class is the simulation setup that runs the whole program. It determines the fog attribute values as well as managing other application-related processes such as adding modules to the application model and connecting application modules in the application model

3.2   Scenarios for Performance Comparison

To investigate the fog's scalability, both scaling out and scaling up approaches are simulated in a smart surveillance environment. Both of the approaches have similar settings in the cloud, proxy and end devices except the fogs. It is worth noting that the proxy here is the Internet Service Provider proxy. Additionally, the fogs operate on three areas where each area has four cameras. These cameras run the same application on both approaches where the application modules include object detector, motion detector, object tracker and a user interface. Scenario 1 represents the scaling out approach where it shows clusters of fogs with similar specifications and scenario 2 presents the scaling up approach consisting of high-end fogs of greater capabilities than the fogs in the scaling out approach. These are illustrated in Fig. 2.

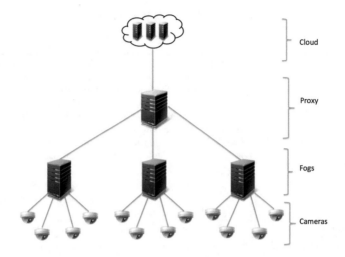

Fig. 2. Generic fog deployment scenario.

3.3 Parameter Configurations

Table 3 shows the default configurations for the cloud, proxy server, fog, and end devices in terms of million instructions per second (MIPS), RAM in gigabyte, bandwidths, hierarchy level, rate per MIPS, and busy and idle powers. There are five configurations for both approaches, named Config 1, Config 2, Config 3, Config 4 and Config 5. In the scaling out approach, each area would have one, two, three, four and five fogs in Config 1, Config 2, Config 3, Config 4 and Config 5, respectively. In scaling out approach, one high-end fog is allocated in all five configurations. However, the high-end fog's specifications are increased to match the total specifications of the fog clusters in the scaling out approach. For instance, when the total RAM is 4 GB (from one fog of 4 GB RAM), 8 GB (from two fogs of 4 GB RAM), 12 GB (from three fogs of 4 GB RAM), 16 GB (from four fogs of 4 GB RAM) and 20 GB (from five fogs of 4 GB RAM) for Config 1, Config 2, Config 3, Config 4 and Config 5 in scaling out approach, the RAM of high-end fog in scaling up approach is similarly set to 4 GB, 8 GB, 12 GB, 16 GB and 20 GB in Config 1, Config 2, Config 3, Config 4 and Config 5. The latencies between the source and destination in our simulation setup are also using the iFogSim's default values where the cloud and proxy connection has a latency of 100ms, proxy to fog and fog to the end devices both have 2ms latency.

4 Results and Discussions

As the accumulated resources of the fogs increases, the cost of execution decreases in both scaling out and scaling up approaches. Our initial results depicted in Fig. 3 has shown that the cloud execution cost is reduced from 61778 to 58891 for the scaling out approach, and from 62476 to 59090 for the scaling

Table 3. Default entity configurations in iFogSim.

Attribute	Cloud	Proxy	Fog	Camera
MIPS	44800	2800	2800	1500
RAM (GB)	40	4	4	2
Uplink bandwidth	100	10000	10000	10000
Downlink bandwidth	10000	10000	10000	10000
Hierarchy level	0	1	2	3
Rate per MIPS	0.01	0	0	0

up approach. The reductions in both approaches is accredited from the shift of application or task processing from the cloud to the fog layer. Although the scaling up approach has exhibited a higher cloud execution cost compared to scaling out approach, the differences are insignificant. Table 4 shows that using the scaling up approach only made less than %2 increase of execution cost in all of the configurations. Nonetheless, both approaches have their own advantages and disadvantages as mentioned in Sect. 2 that still have to be considered.

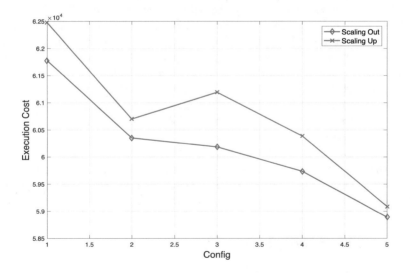

Fig. 3. Execution cost comparison between scaling out and scaling up.

Table 4. Percentage increase of cost execution using the high-end fog.

Config	1	2	3	4	5
Execution cost (%)	+1.10	+0.58	+1.80	+1.1	+0.34

Furthermore, although scaling out has proven to be beneficial in various ways, there are issues in implementing the scaling out approach that are unique in the fog domain as elaborated as follows:

- Availability: Scaling out in a highly distributed and heterogeneous fog environment means that availability will be a challenge. Availability can be affected by failure of hardware, software and internet connectivity. As fog facility can also be included within the already deployed equipment such as in routers, switches, and optical network units [6,13] where they are closer to the end users, hence availability must be guaranteed. Moreover, the fog has a better mobility support as compared to the cloud. Integrating mobile fogs would create a dynamically changing topology, thus affecting the availability. Other factors that would affect the availability include physical threats such as human-made actions either deliberate or unintentional, and natural disasters. Therefore to increase availability, apart from having redundant power supply, multihoming approach can be applied with each of the fogs under both approaches. However, this can be achieved at the price of increasing the cost.
- Security: Ensuring homogeneous security in a heterogeneous fog environment using the scaling out approach would be a necessity. The various kinds of nodes that are added to scale up the network would often have different security requirements and platforms used. This would give rise to incompatibility problems. For each deployment location, it is imperative to understand all kinds of security threats that would occur in many angles as well. Some of the threats include network intrusion, information leakage by attackers during routing, and denial of service. Furthermore, authorization issues would arise as fog resources would be shared by various users. Without proper security measures, the network will be vulnerable to myriad of threats and become easily compromised.
- Network bandwidth: While shifting the processing from cloud to fog would reduce the network bandwidth usage, scaling out the fog domain would imply the addition of more nodes to the fog. Hence, careful planning is required as this approach could introduce network communication overhead as well.

5   Conclusions and Future Work

This paper has demonstrated the different implementations of fog towards the cloud's execution cost. The scaling out presents a cluster of fogs with lower specifications and the scaling up presents a high-end fog. The results show that implementing fogs using the scaling out approach performs slightly better compared to fogs in scaling up approach. Due to time constraints and the limitations of iFogSim, other performance metrics such as throughput, latency and packet delivery ratio cannot be extracted. However, iFogSim can be further modified to provide these metrics to give better assessment of the fog. Nonetheless, this study has shed some insightful directions to work on in the future:

- Load-balancing policy: Since only one application runs in this study, more applications can be added to see how the overall performance will be affected. Hence, load balancing policy needs to be created and applied.
- Mobility: Unlike cloud, fogs support mobility. While mobile fogs add more complexity to the environment, applying mobility allows us to understand how task allocation takes place in a dynamic environment.
- Dynamic clustering: Currently, the iFogSim only allows static fog clusters that have similar attribute values, future work will incorporate dynamic fog clustering where different cluster would have different values.
- Security: Considering that security solutions made for cloud may not suit well in fog, in addition to fogs' mobile capability, fogs are vulnerable in many ways. Although grouping the fogs helps to reduce the overhead and ease the management, it does not guarantee the security of the fog as a whole. Hence, there is a need to evaluate the trustworthiness of the fog layer to ensure that they are able to perform up to expectation.

References

1. Agrawal, D., El Abbadi, A., Das, S., Elmore, A.J.: Database scalability, elasticity, and autonomy in the cloud. In: Lecture Notes in Computer Science (including subseries Lecture Notes in Artificial Intelligence and Lecture Notes in Bioinformatics). LNCS(PART 1), vol. 6587, pp. 2–15 (2011)
2. Appuswamy, R., Gkantsidis, C., Narayanan, D., Hodson, O., Rowstron, A.: Scale-up vs scale-out for Hadoop, pp. 1–13 (2013). http://dl.acm.org/citation.cfm?doid=2523616.2523629
3. Barroso, L., Dean, J., Holzle, U.: Web search for a planet: the google cluster architecture. IEEE Micro **23**(2), 22–28 (2003). http://ieeexplore.ieee.org/xpls/absall.jsp?arnumber=1196112
4. Bonomi, F., Milito, R., Zhu, J., Addepalli, S.: Fog computing and its role in the internet of things. In: Proceedings of the First Edition of the MCC Workshop on Mobile Cloud Computing, pp. 13–16 (2012)
5. Cisco: Fog Computing and the Internet of Things: Extend the Cloud to Where the Things Are, pp. 1–6 (2015). http://www.cisco.com/c/dam/en_us/solutions/trends/iot/docs/computing-overview.pdf
6. Dastjerdi, A.V., Gupta, H., Calheiros, R.N., Ghosh, S.K., Buyya, R.: Fog computing: principles, architectures, and applications. In: Internet of Things: Principles and Paradigms (2016)
7. Gupta, H., Dastjerdi, A.V., Ghosh, S.K., Buyya, R.: iFogSim: a toolkit for modeling and simulation of resource management techniques in internet of things, edge and fog computing environments, pp. 1–22 (2016). http://arxiv.org/abs/1606.02007
8. Hong, K., Lillethun, D.: Mobile fog: a programming model for large-scale applications on the internet of things, pp. 15–20 (2013). http://dl.acm.org/citation.cfm?id=2491270
9. Lopes, M.M., Capretz, M.A.M., Bittencourt, L.F.: MyiFogSim: a simulator for virtual machine migration in fog computing. In: Proceedings of the 10th International Conference on Utility and Cloud Computing, pp. 47–52 (2017)

10. Mahmoud, M.M., Rodrigues, J.J., Saleem, K., Al-Muhtadi, J., Kumar, N., Korotaev, V.: Towards energy-aware fog-enabled cloud of things for healthcare. Comput. Electr. Eng. **67**, 58–69 (2018). https://doi.org/10.1016/j.compeleceng.2018.02.047
11. Mahmud, R., Buyya, R.: Modelling and simulation of fog and edge computing environments using iFogSim toolkit. In: Fog and Edge Computing: Principles and Paradigms, chap. 17, pp. 1–35. Wiley Press (2018)
12. Mahmud, R., Ramamohanarao, K., Buyya, R.: Latency-aware application module management for fog computing environments. ACM Trans. Embed. Comput. Syst. **9**(4), 1–21 (2017)
13. Newaz, S.S., Lee, G.M., Uddin, M.R., Mohammed, A.F.Y., Choi, J.K., et al.: Towards realizing the importance of placing fog computing facilities at the central office of a pon. In: Proceedings of the 19th International Conference on Advanced Communication Technology (ICACT), pp. 152–157 (2017)
14. OpenFog Consortium: Out of the Fog: Use Case Scenarios, pp. 0–9 (2016)
15. OpenFog Consortium: Out of the Fog: Use Case Scenarios, pp. 1–8 (2017)
16. Skala, K., Davidovic, D., Afgan, E., Sojat, Z.: Scalable distributed computing hierarchy: cloud, fog and dew computing. Open J. Cloud Comput. **2**(1), 16–24 (2015)
17. Stojmenovic, I., Wen, S.: The fog computing paradigm: scenarios and security issues, vol. 2, pp. 1–8 (2014). https://fedcsis.org/proceedings/2014/drp/503.html
18. Taneja, M., Davy, A.: Resource aware placement of IoT application modules in Fog-Cloud Computing Paradigm, pp. 1222–1228 (2017)
19. Westbase Technology: Fog Computing vs Cloud Computing: What's the difference? (2017). http://www.westbaseuk.com/news/fog-computing-vs-cloud-computing-whats-the-difference/
20. Ye, D., Wu, M., Tang, S., Yu, R.: Scalable fog computing with service offloading in bus networks. In: Proceeding of the 2016 IEEE 3rd International Conference on Cyber Security and Cloud Computing (CSCloud), pp. 247–251 (2016). http://ieeexplore.ieee.org/lpdocs/epic03/wrapper.htm?arnumber=7545926

Load Balancing in Mobile Cloud Computing Using Bin Packing's First Fit Decreasing Method

P. Herbert Raj[1(✉)], P. Ravi Kumar[2], and P. Jelciana[3]

[1] School of Information and Communication Technology, IBTE SB Campus,
Bandar Seri Begawan, Brunei
herbert.raj@ibte.edu.bn
[2] School of Computing and Informatics, Universiti Teknologi Brunei,
Bandar Seri Begawan, Brunei
ravi2266@gmail.com
[3] Bandar Seri Begawan, Brunei

Abstract. Mobile Cloud Computing (MCC) is the brainchild of the technological revolution of Cloud Computing (CC) and Mobile Computing (MC) with the support of wireless networks, which enables the mobile application developers can create platform independent mobile applications for the users. Cloud Computing is the base for Mobile Cloud Computing to distribute its tasks among various mobile applications. Due to the rapid growth of mobile and wireless devices, it has been a highly challenging mission to send/receive data to mobile devices and accessing cloud computing amenities. In order to overcome the issues in Mobile Cloud Computing such as Low Bandwidth, Heterogeneity, Availability, QoS etc., some new techniques have been implemented so far. One of the core major issues in MCC is load balancing. To address the under-utilization and over-utilization of the processors in MCC, dynamic load balancing techniques plays a key role. In this paper, a new offline load balancing approach is proposed to handle resources in mobile cloud computing. This paper also compares the current approaches of load balancing techniques in MCC.

Keywords: Cloud computing · Mobile cloud computing · Quality of service
Distributed systems · Service level agreement and index name server

1 Introduction

Mobile Cloud Computing is a cutting-edge technology that slowly started dominating the world of wireless technology. The goal of MCC is to deliver mobile users with enriched resources, for instance, prolonged battery life, computation time, communication etc. [1]. The mobile users are receiving tremendous services from the Mobile Cloud Computing such as accessibility to various shared resources, high storage capacity and data integration. The mobile users, small and medium-sized companies

---

P. Jelciana—IT Consultant, Brunei

© Springer Nature Switzerland AG 2019
S. Omar et al. (Eds.): CIIS 2018, AISC 888, pp. 97–106, 2019.
https://doi.org/10.1007/978-3-030-03302-6_9

are benefited by this technology by accessing their data from anywhere, at any time. They enjoyed the secured storage proficiency in clouds without investing in buying new hardware for storage. Lot many mobile users started accessing clouds to process their data in an economical way. So providing the guaranteed service to everyone is the major issues in MCC. To manipulate the multifaceted customers' demands, different services from numerous service providers can be integrated [2]. An "always-on" connectivity is indispensable to the MCC [3]. This type of connectivity can be implemented without much hitches in cloud computing. But, in MCC, it's a quite challenging task. The common problems in networks are congestion, disconnection, network failure and out-of-range. Hence, it is necessary to deploy a dynamic optimized routing algorithm to route the traffic in an effective way without network congestion and other major issues.

In Mobile Cloud Computing, mobile devices offload their full or partial computations to the data centers to ease their computational overhead. The offloading the computations to the cloud will lead to the response latency problem [4]. Hence, it should be handed over to the local clouds, which have fewer resources for the computation. So effectively manage these limited resources is a complicated task in MCC.

The objective of this research paper is to devise a load balancing algorithm that will dispense the vibrant jobs equally to all the processors to achieve beneficial processor utilization and reduce response time. Each and every time, the few VMs are overloaded and VMs are under loaded with dynamic load to process, the load balancing algorithms are used to balance the load with optimal number of processors.

1.1   Load Balancing in Mobile Cloud Computing

Load balancing technique is used to lessen the burden of computationally overloaded data centers. This is a very reliable method to share the load across the datacenter's infrastructures [5]. This method can be classified into Static and Dynamic Load Balancing Algorithms [6]. The following Fig. 1 depicts the classification of load balancing algorithms in a broad manner.

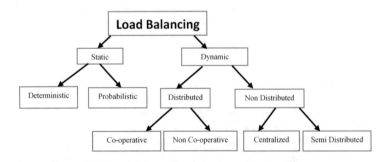

Fig. 1. Classification of load balancing algorithms

a. **Static load balancing algorithms:** This algorithm is very comfortable to deploy. This works well in the homogenous environments with high communication speeds

[7]. It does not monitor the network nodes constantly. It can perform well if the load variation is minimal to VMs. It may not perform well in dynamic cloud environment.

b. **Dynamic load balancing:** In dynamic load balancing, the load is distributed at runtime. This distribution is performed with the current status of VMs. This will help in reduction of communication delay and execution time in the Mobile Cloud environment [5]. There are two types of Dynamic load balancing:

   i *Centralized approach:* In this method, only one node is liable for load distribution and management and all the other nodes are not accountable.

   ii *Distributed approach:* In this method, each node is independent. Every node creates its own database about the neighboring nodes. All decisions are made according to the database information of the nodes [8].

There are numerous benefits of applying load balancing technique in Mobile Cloud environment, such as throughput, minimizing energy consumption and utilization of resources, reduce cost, saving processing time, scalability, fault tolerance and minimizing migration time. Whenever a data center is overloaded with computational jobs, a load balancing algorithm migrates the task to the less utilized datacenter. Also, load balancing boosts the productivity of a data center although upholding the Service Level Agreement (SLA).

2   Background and Related Work

An optimal resource utilization can be achieved by dispensing network load across multiple computers in the network, this procedure is called load balancing. The load balancing algorithms are categorized into two broad methods such as static and dynamic [9]. Static There are lot of pre-computations necessary for static load balancing. Normally, it is defined in the system design or implementation. Hence, it is meant for a simple network with less processing information. Hence, it arises some problems with static load balancing. It can be used in some networks where there is less information to process. The dynamic load balancing algorithms are used in more complex networks. The routing and balancing decision will be changed according to the lot of runtime properties in the network. The current state of the system will be taken into consideration for load balancing decisions. There are few load balancing algorithms which are discussed in this section.

2.1   Load Balancing Algorithms

In the process of information retrieval, data duplication will be the major issue. In order to minimize the problem, INS (Index Name Server) Scheme [10] is proposed. The numerous parameters in this process will decide the optimization selection point in this method. There is a major disadvantage of this method, it lacks in predicting the forthcoming behaviour.

A new algorithm is proposed for the purpose of load balancing known as CLBDM (Central Load Balancing Decision Model) [11]. Round Robin [12] is the fairly renowned algorithm for load balancing. In this method, all the processes are divided among each processor. The CLBDM is the sophisticated version of the Round Robin algorithm. In this algorithm, connection time is calculated and if it surpasses the allowed limit, then it will be terminated. This reduces the physical administration. The disadvantage is single point failure.

The swarm intelligence – Ant colony is used for optimal load balancing [13, 14]. The optimal load balancing can be found by generating artificial ants, these ants search the solution space. The exploration of new paths will be in a probabilistic way. The implementation of this algorithm will yield the optimal solution in more complex networks.

Another load balancing algorithm is proposed for the cloud is named as Artificial Bee Algorithm [15]. This algorithm is based on the honeybee's behaviour for gathering nectar. This increases the throughput and reduces the time for a task to wait in a queue.

3   The Proposed Bin Packing Algorithm for Load Balancing

The Fig. 2 depicts the architecture and load balancing method in a cloud by using Bin Packing's First Fit Decreasing method. In this off-line approach, all the jobs are given equal importance and user submitted jobs are arranged in decreasing order using Heap Sorting method. Resources for the underlying networks are provided by the queue manager. In cloud computing, the queue manager is responsible for the utilization of all the resources. It keeps the track of the systems, which are all currently running the jobs by balancing the load among the meta scheduler and its disposal [16]. The load balancer balances the load by using Bin Packing's First Fit Decreasing (FFD) among the resources. The method of load balancing is explained in the Sects. 3.2 and 3.3 in detail.

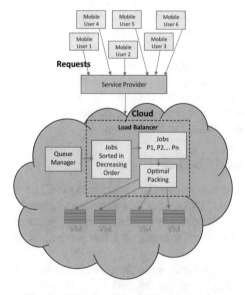

Fig. 2. FFD algorithm for load balancing

3.1 Bin Packing Problem

Bin packing is an NP-Hard combinational optimization problem [17]. The main objective of this method is to pack a set of items into a minimum number of bins. In this problem, n objects that have to be placed in bins of equal capacity L. Object i requires li units of bin capacity. The objective is to determine the minimum number of bins needed to accommodate all n objects. No objects may be placed partly in one bin and partly in another.

> Let L = 7, n = 6 and $(l1,l2,l3,l4,l5,l6)$ = (3,1,6,4,5,2)
> 3 + 1+6 + 4+5 + 2 = 21 divided by capacity 7
> Lower bound = 21/7 = 3
> Available bins are 4 with capacity 7

a. **First Fit Method:** Here, 4 number of bins are required to fit all objects. Moreover, space also wasted in each bin. The above Fig. 3 represents the wasted space in white color. Furthermore, the solution is not optimal.

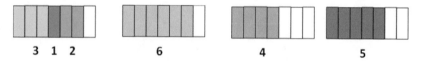

3   1   2   **6**   **4**   **5**

Fig. 3. First fit method

b. **Best Fit and Worst Fit Method:** For the example stated above, the Best Fit and Worst Fit methods are required also the same number of bins to fit all objects.
c. **First Fit, Best Fit and Worst Fit Decreasing Methods:** First Fit Decreasing method is very simple to perform and yields an optimal solution. In the First Fit Decreasing, the objects are ordered in descending $(l1,l2,l3,l4,l5,l6)$ = (6,5,4,3,2,1). If first object fits in first bin it will be assigned else new bin will be selected. In this method, the total number of required bins are only 3 and there is no waste of space in each bin as shown in the Fig. 4.

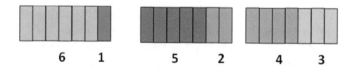

6   1   **5   2**   **4   3**

Fig. 4. First fit decreasing, best fit decreasing and worst fit decreasing methods

d. **Optimal Packing:** In Optimal Bin Packing [18], the feasible objects are chosen to fulfill the bins. In the following Fig. 5, the objects 3 & 4, 5 & 2 and 1 & 6 are combined to fill the bins. In this approach, only 3 bins are required to fill out all the objects and there is no wastage of space.

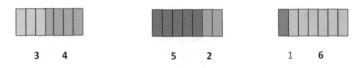

3   4   5   2   1   6

Fig. 5. Optimal packing

3.2   Bin Packing in Mobile Cloud Computing

Simple heuristics for the bin packing problem is given here. These will not, in general, obtain optimal packing. They will, however, obtain packing that uses only a small fraction of bins more than an optimal packing [19].

In this proposed approach, n jobs that have to be placed in processors of equal capacity L. Job i requires li units of processor capacity. The objective is to determine the optimal number of bins needed to accommodate all n jobs. No job may be placed in one processor and partly in another. For example Let L = 100, n = 6, and $(l1,l2,l3,l4, l5,l6)$ = (45, 60, 30, 70, 55, 40). The following figures show the packing of all six jobs in the processors. The optimal solution for this example is three processors. It's shown in the Fig. 6.

1   2   3

Fig. 6. Optimal bin packing

a. **First Fit Method:** The Fig. 7 shows the First Fit method to pack all the six jobs. There are four processors needed to pack all the six jobs.

1   2   3   4

Fig. 7. First fit method

b. **First Fit Decreasing Method:** Since all the decreasing bin packing methods almost yields the same results, the FFD offline method is appropriated for MCC load balancing. The Fig. 8 shows the First Fit Decreasing method to pack all the six jobs. There are only three processors required to pack all the six jobs. This method yields good result in load balancing than the on-line bin packing method in MCC. The available online algorithms do not know the future loads. It's a major drawback in getting an optimal solution. The competitive analysis [20] of few online and offline

1   2   3

Fig. 8. First fit decreasing method

algorithms shown in the following Table 1 with respect to the competitive ratio. The Fig. 9 clearly shows the qualities of the offline algorithms are better than online algorithms.

Table 1. Comparison of on-line and off-line algorithms

Algorithm	Competitive analysis
On-line bin packing	1.58889
First fit decreasing	1.22
Best fit decreasing	1.22
Next fit	2.0
Best fit	1.7
Worst fit	1.7

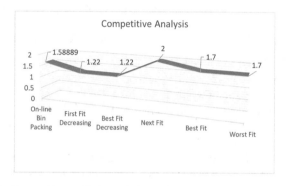

Fig. 9. Competitive analysis variations of on-line and off-line algorithms

3.3   Algorithm

Algorithm code for First Fit Decreasing

```
1. Initialize cloud
2. Initialize load
3. Sort all jobs in decreasing order
4. For all job j = n to 1 do
5. For all processors p = 1, 2…,m do
6. If job j fits in processor p then
7. Assign job i into processor p
8. End if
9. End for
10.If the current job cannot be assigned to any
   processor then
11.job j will be assigned to a new processor
12.End if
```

Step 1 takes $0(n \log n)$ time [21]. In the worst-case, when a new job is inserted the new processor should be assigned. Thus, there are $1, 2, 3, ..., n-1$ executions of the inner loop, which yields an asymptotic factor of $O(n^2)$. The algorithm is obviously dominated by the running time of First-Fit, which yields a factor of $O(n^2)$ [22].

3.4   Theorem

Theorem: The bin packing problem is NP-hard [17].

Proof: *Let* $\{a1, a2..., an\}$ be an instance of the partition problem. Define an instance of the bin packing problem as

$$Li = ai, 1 < = i < = n, \text{ and } L = \sum ai/2 \qquad (1)$$

The minimum number of bins needed is two if and only if there is a partition for $\{a1, a2, ..., an\}$

The following are four simple heuristics for bin packing problem.

a. **First Fit (FF):** Index the processors 1, 2, 3... All processors are initially filled to level zero. Jobs are considered for packing in the order 1, 2, ..., n. To pack job i, find the least index j such that processor j is filled to a level $r, r < = L - li$. Pack i into processor j. Processor j is now filled to level $r + li$ [16].

b. **Fist Fit Decreasing (FFD):** Reorder the jobs so that $li > = li + 1, 1 < n$. Now use FF to pack the jobs [16].

c. **Best Fit (BF):** The preliminary states on the processors and jobs are same like FF. To pack job i, find the least index j such that processor j is filled to a level r, $r <= L - li$ and as large as possible. Pack i into processor j. Processor j is now filled to level $r + li$ [16].

d. **Best Fit Decreasing:** Reorder the jobs so that $li > = li + 1, 1 < n$. Now use BF to pack the jobs [16].

4   Conclusion

Distributing the work into various resources is the major task of the load balancing algorithms in Mobile Cloud Computing. In this research paper, achieving load balancing in mobile cloud computing using bin packing method is proposed. This proposed algorithm tried to balance the load equally among the available processors by using the First Fit Decreasing offline method. This method considerably improves the utilization of the server and reduce the vacant space in a server. So this algorithm intensifies the server utilization and enriches the throughput considerably. This algorithm can also be devised for varying size of the processors in the server. This method efficiently handles the load and balances among available resources. This algorithm is an offline algorithm [20] and with the approximation ratio 1.22. The online algorithms have two shortcomings. One is 'Online Constraint': online algorithms could not predict the upcoming requests. So the forecasting the future items in bin packing and future points in clustering will be impossible. Second is 'Sequential Constraint': online algorithms fabricate their result sequentially. They could not change their previous decision, e.g., if a job placed in a processor, then it cannot be replaced and also randomization will not help significantly. So far no online algorithm has reached less than 1.54037 [20] competitive ratio. This proposed offline FFD algorithm can be modified to achieve better results. In order to extend this research work, researchers can propose some online and offline algorithms to achieve better utilization of resources.

References

1. Herbert Raj, P., Ravi Kumar, P., Jelciana, P.: Mobile cloud computing: a survey on challenges and issues. Int. J. Comput. Sci. Inf. Secur. (IJCSIS) **14**(12), 165–170 (2016)
2. Sarddar, D.: A New Approach on Optimized Routing Technique for Handling Multiple Request from Multiple Devices for Mobile Cloud Computing, vol. 3(8), pp. 50–61, August 2015. ISSN 2321-8363
3. Huerta-Canepa, G., Lee, D.: A virtual cloud computing provider for mobile devices. In: 1st ACM Workshop on Mobile Cloud Computing & Services: Social Networks and Beyond (MCS). ACM, June 2010
4. Wei, X., Fan, J., Lu, Z., Ding, K.: Application scheduling in mobile cloud computing with load balancing. J. Appl. Math. **2013**(409539), 13 p. http://dx.doi.org/10.1155/2013/409539
5. Dhinesh, B.L.D., Krishna, P.V.: Honey bee behavior inspired load balancing of tasks in cloud computing environments. J. Appl. Soft Comput. **13**(5), 2292–2303 (2013)
6. Gabi, D., Ismail, A.S., Zainal, A.: Systematic review on existing load balancing techniques in cloud computing. Int. J. Comput. Appl. (0975–8887) **125**(9) (2015)
7. Singh, A., Juneja, D., Malhotra, M.: Autonomous agent based load balancing algorithm in cloud computing. Procedia Comput. Sci. J. **45**(1), 832–841 (2015)
8. Kaur, R., Luthra, P.: Load balancing in cloud computing. In: Proceedings of International Conference on Recent Trends in Information, Telecommunication and Computing, ITC, Association of Computer Electronics and Electrical Engineers (2014). doi:02.ITC.2014.5.92
9. Anjali, J.G., Singh, M., Singh, C., Sethi, H.: A new approach for dynamic load balancing in cloud computing. IOSR J. Comput. Eng. (IOSR-JCE), 30–36. www.iosrjournals.org, e-ISSN 2278-0661, p-ISSN 2278-8727
10. Wu, T.-Y., Lee, W.-T., Lin, Y.-S., Lin, Y.-S., Chan, H.-L., Huang, J.-S.: Dynamic load balancing mechanism based on cloud storage. In: IEEE International Conference on Computing, Communications and Applications (ComComAp), pp. 102–106, January 2012
11. Radojevic, B., Zagar, M.: Analysis of issues with load balancing algorithms in hosted (cloud) environments. In: 34th IEEE International Convention on MIPRO, pp. 416–420, May 2011
12. Randles, M., Lamb, D., Taleb-Bendiab, A.: A comparative study into distributed load balancing algorithms for cloud computing. In: 24th IEEE International Conference on Advanced Information Networking and Applications Workshops, pp. 551–556 (2010)
13. Rajagopalan, S., Naganathan, E.R., Herbert Raj, P.L.: Ant Colony Optimization Based Congestion Control Algorithm for MPLS Network, vol. 169, pp. 214–223. Springer, Heidelberg (2011). Print ISBN 978-3-642-22576-5, Online ISBN 978-3-642-22577-2
14. Zhang, Z., Zhang, X.: A load balancing mechanism based on ant colony and complex network theory in open cloud computing federation. In: IEEE International Conference on Industrial Mechatronics and Automation (ICIMA), vol. 2, pp. 240–243, May 2010

15. Yao, J., He, J.: Load balancing strategy of cloud computing based on artificial bee algorithm. In: IEEE International Conference on Computing Technology and Information Management (ICCM), vol. 1, pp. 185–189, April 2012
16. Singh, K.: Energy efficient load balancing strategy for mobile cloud computing. Int. J. Comput. Appl. (0975–8887) **132**(15) (2015)
17. Horowitz, E., Sahani, S., Rajasekaran, S.: Fundamental of Computer Algorithms. Galgotia Publications Pvt. Ltd., Delhi (2008)
18. Edexcel Decision Mathematics 1. Packing and searching algorithms, Hegarty. https://hegartymaths.com/, https://www.youtube.com/watch?v=kiMFyTWqLhc
19. Kasmir Raja, S.V., Herbert Raj, P.: Balanced traffic distribution for MPLS using bin packing method. In: 2007 3rd International Conference on Intelligent Sensors, Sensor Networks and Information. IEEE, December 2007. https://doi.org/10.1109/issnip.2007.4496827, ISBN 978-1-4244-1501-4
20. Boyar, J., Kamali, S., Larsen, K.S., Lopez-Ortiz, A.: Online Bin Packing with Advice. Trends in online algorithms, July 2014
21. Iyer, K.V.: Bin packing – an approximation algorithm: how good is the FFD heuristic - a weak bound, April 2008. https://www.nitt.edu/home/academics/departments/cse/faculty/kvi/Bin%20Packing%20FFD%20heuristics.pdf
22. Rieck, B.: Basic Analysis of Bin-Packing Heuristics, Publicado por Interdisciplinary Center for Scientific Computing. Heildelberg University (2010)

Independent and Distributed Access to Encrypted Cloud Databases

Marlapalli Krishna[1], K. Chaitanya Deepthi[1], Soni Lanka[2,3,P],
S. B. P. Rani Bandlamudi[1], and Rama Rao Karri[4(✉)]

[1] Sir C. R. Reddy College of Engineering, Eluru, Andhra Pradesh, India
[2] Department of Computer Sciences and Systems Engineering, Andhra University,
Visakhapatnam, India
karri.sony@gmail.com
[3] Computer Sciences, Faculty of Science, Universiti Brunei Darussalam,
Gadong, Brunei Darussalam
[4] Faculty of Engineering, Universiti Teknologi Brunei, Gadong, Brunei Darussalam
karri.rao@utb.edu.bn

Abstract. The present generation prefers to store their data in the cloud, which brings them mobility and ease of extracting data from any device to anywhere in the world. Since data in the cloud is going to be placed online, it is important that these data in the clouds are well secured. The most important security challenge with data in the clouds is that the client was not aware of where the data is stored. Which may be preyed on by 3rd party clients or attackers. The data should be accessible solely by trustworthy parties that don't embrace cloud suppliers and hence information should be encrypted. We propose Secure DBaaS architecture because this enables cloud tenants to acquire full advantage of DBaaS qualities, like accessibility, scalability and reliability, while not exposing unencrypted information to the cloud supplier.

Keywords: Security · Cloud · Secure DBaaS · Reliability · Privacy

1 Introduction

Cloud computing is another prototype that is built on disseminated computing, parallel and virtualisation, utility processing, and service-oriented outline [1, 2]. In the recent years, distributed computing has developed commonly of the premier powerful ideal models with the information technology [3–6]. Clients/users utilize services while not stressing concerning inbuilt background computation and storage [7–9]. The distributed storage framework is considered as an outsized scale appropriated capacity framework that comprises of the numerous independent storage service providers. Knowledge hardiness might be a noteworthy request for capacity frameworks [10–12]. A technique to create learning strength is to duplicate a message determined each capacity server stores a replica of the message. A "column database management system (CDBMS)", might be a distributed information that conveys "computing as a service instead of a

© Springer Nature Switzerland AG 2019
S. Omar et al. (Eds.): CIIS 2018, AISC 888, pp. 107–116, 2019.
https://doi.org/10.1007/978-3-030-03302-6_10

product". It's the sharing of resources, programming, and information between duplicate gadgets over a system that is, for the most part, the web.

The Secure DBaaS framework is custom-built to cloud platforms and doesn't present any proxy or on the other hand representative between the cloud provider and the customer [13]. Eliminating any trusted intermediate server enables Secure DBaaS to accomplish a similar versatility, reliability and accessibility of a cloud DBaaS. Unlike Secure DBaaS, designs depending on a trusted intermediate proxy [15].

In this research study, we offer a Secure DBaaS as the primary arrangement that allows cloud tenants to have full advantage of DBaaS characteristics, such as, accessibility, scalability and reliability. Our study aims to establish cloud data administrations with learning secrecy and along these lines the likelihood of the capital punishment co-happening operations on encrypted information. We tend to utilise the cloud for transferring proprietor's information. This encrypted information is kept at the cloud alongside its encoded data. This is an essential solution supporting topographically conveyed buyers to connect on to relate encoded cloud data and to execute co-happening and independent tasks together with those adjusting the data structure.

2   System Overview

The layout is set up by a "database administrator (DBA), and the execution of SQL" activities are depicted on encoded information through two stages. Primarily, "a naïve setting described by a solitary customer, and secondary through reasonable settings where the database services are accessed by simultaneous customers". The system is designed on the framework of integration of Cloud database, Meta-data management and Encrypted algorithm [13, 14].

Cloud Database: In this, that tenant information area unit protected is assumed to be in an online database. The confidentiality of the information has to be preserved and also the information structure which is a result of row/column names. This could yield info concerning saved information. We tend to differentiate the methods for encoding the information structures and the client information.

Metadata Management: Data created by a Secure DBaaS contain every one of the information that is important to manage SQL proclamations over the encoded data in a surpassing technique clear to the client. "Metadata management" represents a creative plan as a result of Secure DBaaS is that the initial design is storing all metadata inside the untrusted cloud data alongside the scrambled tenant data.

Encrypted Algorithm: The encoding algorithms are selected which are accustomed to encipher and decipher all the information is stored in an encoded text.

3   System Design

Secure DBaaS is intended to enable different independent clients to interface straightforwardly to the "untrusted cloud DBaaS with no intermediate server". Figure 1 depicts the general architecture of Secure DBaaS. We tend to accept that "an occupant association gets a cloud database benefit from an untrusted DBaaS provider". The occupant at that point deploys one or a considerable measure of machines (Client one through N) and introduces a Secure DBaaS relation on each one of them [13]. This relation allows a client to "connect to the cloud DBaaS to administer it", to check and compose information. Secure DBaaS is proposed to allow numerous clients to join on to the "untrusted cloud DBaaS with no intermediate server".

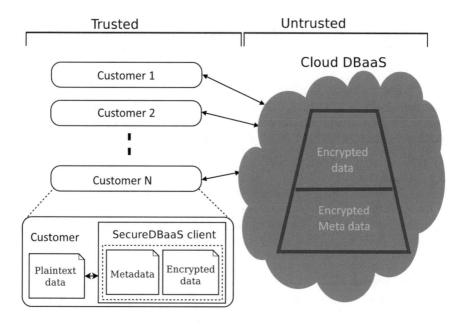

Fig. 1. The general architecture of Secure DBaaS

3.1   Cloud Database

We tend to accept that tenant information zone unit saved in a database. We must safeguard the classification of the data and even its structure because the row/column segments could yield information. The ways for encoding the information structures and the tenant data should be distinguished.

3.2   Metadata Management

Information generated by a Secure DBaaS comprises of knowledge that is important to oversee SQL queries over the encoded data. Metadata administration represents a guileless arrangement because of Secure DBaaS which underlie the outline of all the metadata inside the untrusted cloud data adjacent to the encrypted tenant data.

3.3   Encryption Algorithm

Selecting the coding algorithms accustomed cypher and decode all the info stored in the information table. "Secure DBaaS is intended to permit various and independent customers to interface straightforwardly to the untrusted cloud DBaaS with no intermediate server".

3.3.1   Database Creation

In this module, the consumer generates its data and store information within the kind or columns and rows. After creation of data, the consumer conjointly creates its data which can facilitate later communication rather than whole information.

3.3.2   Choice of Secret Writing and Coding Rule

In this module, the secret writing rule to inscribe and rewrite the generated data and its corresponding meta-data are chosen. It'll give security to whole information of consumer that is to be uploaded to the cloud server.

3.3.3   Cloud Database

Cloud data is that the administration provider which gives administrations to the tenants. All the scrambled data from data proprietor is transferred to the cloud that gives synchronous access to cloud dB to the topographically deployed purchasers. Cloud dB contains encoded data and its scrambled information.

3.3.4   Application

The application of this system to the cloud is made in this phase. However, we are going to apply these all to other cloud management as well. We tend to use the key to access cloud information, where data is uploaded to cloud. Primarily, if our key is correct, we are going to get encoded information and then by random coding keys, we will achieve the ultimate output of the plaintext information. Here the user is provided with the necessary input within the type of SQL query. However, the consumer can produce relevant information which is entered in rows of information. Afterwards, the database is created. The secret writing rule is implemented on the information and its data. The final output provides the encoded information with all its necessary info and relevant keys to be used.

4 Implementation

This stage is performed in three different phases, namely, Knowledge Management, Data Management and finally relevant algorithms. These phases and applications are details as follows:

4.1 Knowledge Management

Cloud information acts as service supplier for clients. All the information and knowledge are stored within the electronic service provider. Thus for making data segregated into rows/columns which can be easily accessible using the appropriate SQL query.

4.2 Data Management

Data management ways represent an imaginative plan as a result of Secure DBaaS is that the 1st design is storing all data within the untrusted cloud information beside the encrypted tenant knowledge.

Secure DBaaS uses two styles of data. Namely;

- Information data area unit regarding the entire information. There's solely one instance of this data sort for every information.
- Table data area unit related to one secure table.

This style selection makes it double to spot that data that is needed to implement any necessary SQL statement, so a secure DBaaS customer has to fetch solely the data that is concerned with the SQL query. This style alternative minimises the quantity of data that every Secure DBaaS client needs to extract data from the untrusted cloud information, therefore reducing information measure consumption and time interval. Furthermore, it allows multiple clients to access several data. Database metadata contain the cryptography keys that area unit used for the protected varieties. A particular cryptography key's associated with all the attainable combos of knowledge kind and cryptography kind. Hence, the information data represent a hoop and don't contain any info regarding tenant information. The structure of a table meta-data is depicted in Fig. 2.

Every column data in this meta-data table contains the following structure as shown in Fig. 2. The segregation and purpose of each row are explained as:

- Plain name: In this row, "the name of the corresponding column of the plaintext" is stored.
- Encrypted name: This row stores "the name of the column of the secured table". This can be the sole info that integrates a "column to the corresponding plaintext column" as a result of column names of the secure table.
- Cryptography key: This plays a significant role in securing the data in the meta-data table. This key is utilised to encipher and decode all the info kept within the column.
- Secure type: This is considered as the secure sort of the meta-data table. This enables Secure DBaaS clients to be told regarding the type of data and also the cryptography policies related to a column.

Secure DBaaS stores data within the data storage table that's situated within the untrusted cloud as the information. This can be an ingenious alternative that augments flexibility, however, opens two novel problems regarding efficient information retrieval and information confidentiality. To permit Secure DBaaS shoppers to control data through SQL statements, we tend to save information and table data during a tabular type. Indeed, even metadata secrecy is ensured through encryption. Furthermore, Secure DBaaS clients can utilise storing strategies to reduce the bandwidth overhead.

4.3 Algorithms

Encryption algorithms or applied to cypher the information. There are various secret writing algorithms symmetric and uneven. However, we are going to apply the rhombohedral rule that proved to be a key distribution to all clients, and there will be no different non-public key for each user.

5   Experimental Results

The correctness of Secure DBaaS to various DBaaS cloud arrangements is demonstrated by "implementing and handling encrypted database activities on real and emulated cloud frameworks". This Secure DBaaS model supports MySql based databases. It was observed that while porting "Secure DBaaS to various DBMS, required minor alterations which are related to database connector" and very insignificant alterations to the codebase.

To assess "encryption costs, the customer measures the execution time" of the proposed SQL sequences of the TPC-C. Encryption times required for these benchmark operations are presented in the Fig. 3. TPC-C activities are assembled based on the class

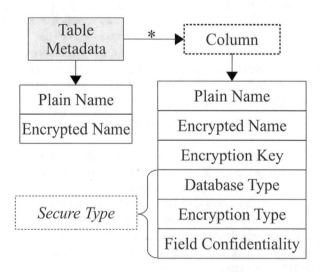

Fig. 2. Schematic representation of metadata table.

of exchange. From this figure, it was observed that the encryption time is lower than 0.1 ms for the operations except for only two instances. These two cases are an exception for the two tasks of Stock Level and Payment transactions. This deviation can be due to the utilisation of the preserving encryption that is vital for regular inquiries.

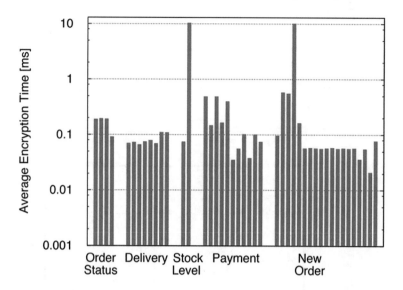

Fig. 3. Results depicting the encryption times of benchmark operations

To assess the performance of encoded SQL tasks, the most commonly executed tasks like "INSERT, SELECT, DELETE and UPDATE", commands of the TPC-C benchmark are chosen. The response times of these tasks are shown in Fig. 4(a) and (b). It was observed that the response times of Secure DBaaS SQL commands for tasks like "SELECT, UPDATE, and DELETE", are almost doubled. Whereas for the INSERT task, it accomplishes in triple response time when compared to the plain text. This higher response time can be because an "INSERT" command needs to scramble all columns.

The throughput for increase number independent clients with respect to 40 and 80 ms are shown in Fig. 5(a) and (b). These measures are "optimistic representations of continental and intercontinental delays". The "patterns of the Secure DBaaS lines are nearly close to those of the TPC-C database, thus signifying that Secure DBaaS encoded database does not influence versatility for a plain database". Also, the system characteristics tend to cover "cryptographic overheads for any number of customers". This outcome is essential since it affirms that "Secure DBaaS is a substantial and handy solution for ensuring data confidentiality in real cloud database services".

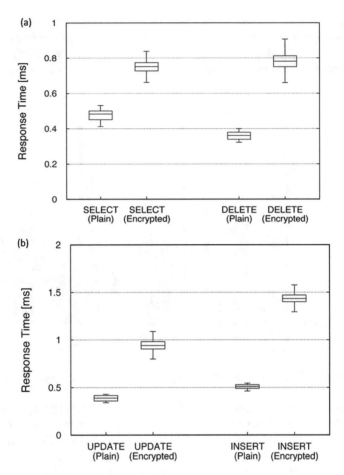

Fig. 4. Response time for operations like plain and encrypted operations for tasks (a) Select & Delete (b) Plain & Insert

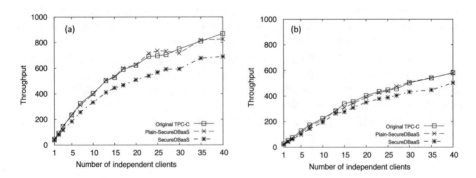

Fig. 5. Profiles depicting the throughput against a number of independent clients for (a) 40 ms and (b) 80 ms

6   Conclusion

In this research study, the innovative architectural design is proposed which ensures the privacy of the data stored in the public cloud. Unlike other methodologies, our method does not depend on any intermediate proxy, which is the bottleneck restricting the versatility of typical cloud database management. The outcomes of this study from experimental results indicate that the data encryption performance on the response time is very insignificant. This outcome is significant since it affirms that "Secure DBaaS is a substantial and handy solution for ensuring data confidentiality in real cloud database services".

References

1. Wu, D., Rosen, D.W., Wang, L., Schaefer, D.: Cloud-based design and manufacturing: a new paradigm in digital manufacturing and design innovation. Comput.-Aided Des. **59**, 1–14 (2015)
2. Armbrust, M., Fox, A., Griffith, R., Joseph, A.D., Katz, R., Konwinski, A., Lee, G., Patterson, D., Rabkin, A., Stoica, I., Zaharia, M.: A view of cloud computing. Commun. ACM **53**(4), 50–58 (2010)
3. Venkata, R.N., Nagesh, P., Soni, L., Karri, R.R.: Big Data Analytics and IoT gadgets for Tech savvy Cities. Advances in Intelligent Systems and Computing (2018)
4. Chang, V., Kuo, Y.H., Ramachandran, M.: Cloud computing adoption framework: a security framework for business clouds. Futur. Gener. Comput. Syst. **57**, 24–41 (2016)
5. Madhavi, R., Karri, R.R., Sankar, D.S., Nagesh, P., Lakshminarayana, V.: Nature inspired techniques to solve complex engineering problems. J. Ind. Pollut. Control **33**(1), 1304–1311 (2017)
6. Lanka, S., Karri, R.R., Prasad, R.P.V.G.D., Peri, S.R.: Data mining, image processing and pattern recognition for bio-medical informatics. In: 4th Biotechnology World Congress, vol. 1(01), pp. 214–215 (2016)
7. Venkata, R.N., Seravana, K.P.V.M., Puvvada, N.: Analytic architecture to overcome real time traffic control as an intelligent transportation system using big data. Int. J. Eng. Technol. **7**(2.18), 7–11 (2018)
8. Venkata, R.N., Puvvada, N., Seravana, K.P.V.M., Vignesh, U.: IoT Based Scientific design to conquer constant movement control as a canny transportation framework utilizing huge information available in Cloud Networks. J. Adv. Res. Dyn. Control. Syst. **10**(07-Special Issue) (2018)
9. Ali, M., Samee, U.K., Athanasios, V.V.: Security in cloud computing: opportunities and challenges. Inf. Sci. **305**, 357–383 (2015)
10. Kevin, H., Murat, K., Latifur, K., Bhavani, T.: Security issues for cloud computing. Int. J. Inf. Secur. Priv. **4**(2), 39–51 (2010)
11. Lanka, S., Madhavi, M.R., Abusahmin, B.S., Puvvada, N., Lakshminarayana, V.: Predictive data mining techniques for management of high dimensional big-data. J. Ind. Pollut. Control **33**, 1430–1436 (2017)
12. Karri, R.R., Sahu, J.N.: Modeling and optimization by particle swarm embedded neural network for adsorption of zinc (II) by palm kernel shell based activated carbon from aqueous environment. J. Environ. Manage. **206**, 178–191 (2018)

13. Luca, F., Michele, C., Mirco, M.: Distributed, concurrent, and independent access to encrypted cloud databases. IEEE Trans. Parallel Distrib. Syst. **25**(2) (2014)
14. Karri, R.R., Venkateswarlu, Ch.: Nonlinear model based control of complex dynamic chemical systems. J. Adv. Chem. Eng. Res. **2**(1), 1–19 (2013)
15. Gentry, C.: Fully homomorphic encryption using ideal lattices. In: 41st Annual ACM Symposium on Theory of Computing, May 2009

Background Traffic Load Aware Video Class-Lecture Client Admission in a Bandwidth Constrained Campus Network

Siti Aisyah binti Haji Jalil[1], Amirah Ruyaieda binti Haji Abdul Majid[1],
Mohammad Ramzi bin Salleh[1], Fatin Hamadah Rahman[1],
Mohammad Rakib Uddin[1], S. H. Shah Newaz[1,2(✉)], and Gyu Myoung Lee[3]

[1] Universiti Teknologi Brunei, Jalan Tungku Link, Gadong BE1410,
Brunei Darussalam
{bcns0314.021,bcns0314.014,bcns0314.010,P20171005}@student.utb.edu.bn
[2] KAIST Institute for Information Technology Convergence, 291 Daehak-ro,
Yuseong-gu, Daejeon 34141, South Korea
{rakib.uddin,shah.newaz}@utb.edu.bn
[3] Faculty of Engineering and Technology, Liverpool John Moores University,
Liverpool, UK
G.M.Lee@ljmu.ac.uk

Abstract. Video class-lecture streaming is regarded as a popular means in improving quality of teaching and learning in schools and universities. Several research findings reveal that, recorded lecture videos (streamed over the Internet) are a useful supplement to non-classroom learning. Despite knowing this importance, some schools or universities are reluctant to use video lecture streaming service in their campus network, thinking video streaming service would impose additional traffic load in their network. In fact, in a bandwidth constrained campus network, other regular traffic flows may experience lower throughput, packet drop and delay due to presence of class lecture video streaming traffic, resulting in deteriorating Quality of Experience (QoE) of campus users. In this paper, we propose video streaming service model for the bandwidth constrained campus networks. We refer to our solution as Class Lecture on Demand (CLD) service that can be easily adopted in a campus. CLD defines the policies for admitting number of clients that request for video streaming service taking into account peak hour and off-peak hour background traffic load. This paper provides a detailed procedures, showing how a network administrator in a bandwidth constrained campus network can measure the maximum number class lecture streaming requests that a video streaming server should accommodated at different part of a day without affecting other traffic flows. Additionally, we provide an insightful discussion (policies) in order to make video lecture streaming in a bandwidth constrained campus network easily adopted.

Keywords: Captured class-lecture · Video streaming policy · QoE
Background traffic · Campus network

© Springer Nature Switzerland AG 2019
S. Omar et al. (Eds.): CIIS 2018, AISC 888, pp. 117–128, 2019.
https://doi.org/10.1007/978-3-030-03302-6_11

1 Introduction

Class-lecture video streaming could effect attendance and engagement in a class. Further, while a lecturer delivers a lecture, the video capturing process may restrict his/her spontaneous lecture-delivery style [2]. Despite these downsides, class-lecture video streaming is found to be an effective mean to improve learning outcomes [3]. Technological advancement for multimedia content generation, processing and storage is contributing rapid growth of video lecture service. Aside from this, we believe, one important factor that is playing important role to facilitate this move is availability of hand-held devices with low cost and high processing capability.

Authors in [2] highlight several important aspects that may impede adopting class-lecture video streaming. In our opinion, due to limited network resources some schools and universities may reluctant as well to introduce this service to their students. The main objective of this paper is to propose a class-lecture video steaming service model over a bandwidth constrained campus networks. We propose this model taking into account: *(i)* university campus regular traffic load (without class-lecture streaming) throughout a day, *(ii)* number of students (clients) interested in the video streaming service, and *(iii)* Quality of Experience (QoE) of clients. We refer to our solution as Class Lecture on Demand (CLD) service. In this paper, the prime objective of CLD is to introduce a set of policies in order to facilitate video class-lecture streaming in a bandwidth constrained campus network. Following the policies, we provide based on our study, one can find the maximum number of video class-lecture streaming clients that can be accommodated with providing satisfactory QoE during peak and off-peak hours of a day (without deteriorating the existing QoE of background traffic flows).

The remainder of this paper is organized as follows. Section 2 discusses relevant research. Section 3 presents details of video lecture streaming performance study in a bandwidth constrained campus scenario. Section 4 provides recommendations for further improvement of video streaming service in a bandwidth constrained campus network. Section 5 concludes this paper.

2 Related Work

2.1 Research on QoE Evaluation

Research activities on the QoE evaluation of video streaming has been increasing in the recent years. To date, there are many research efforts to understand the influence of packet delay, packet drop and jitter in video streaming service on QoE of end users. For instance, Bhamidipati and Kilari [1] had made a survey on 41 users to analyse the quality of the video shown to them with varying delays. Based on their research, they concluded that the delay variation contributes tremendously in degrading users' QoE (a small delay variation can have catastrophic influence on video quality). A similar study was conducted in [7] in an attempt to reveal how the network QoS can influence the QoE of video streaming over HTTP. From their findings, they concluded that network throughput may drop due to packet loss and delay.

Zhengyou et.al. in [8] used a QoS/QoE mapping assessment model for the relationship between QoS and QoE. They conducted an experiment using NS2 network simulator and *myEvalvid* in order to meet their research objectives. Their findings can be used to predict QoE based on QoS parameters. Authors in [4] provided as well a correlation model for QoE evaluation in IPTV.

In [5], authors observed how packet loss may influence video stream transmission. They emphasized that a larger packet size will lead to better quality of received video than that of smaller packets carrying video frames (larger the packet size, the better the quality of video received). The authors conclude that, if a video is segmented into smaller packets, it is possible that the video may suffer from losses of I-frames. Note that, I-frames is increasingly important to decode the frame, and hence loss of I-frames will lead to affecting the entire groups of frames.

There are two popularly used methods to estimate the QoE: objective assessment and subjective assessment. In objective assessment approach, mathematical modeling is required in order to infer human perception relating to a video quality. As this approach relies solely mathematical formulation, this approach is cheap and quick. On the other hand, in subjective assessment method, a subject is shown a video and then, the opinion of the subject relating to the perception on quality of the video is recorded. That is, subjective method is a approach for measuring subject's perceived opinion on the quality of a video—it is the analysis of the subjects' Mean Opinion Score (MOS). To date, there are several objective assessment methods have been introduced already. In fact, one objective assessment method (mathematical model) can be regarded as a reliable model if the objective assessment results are close to a subjective assessment result based on the same video. Interestingly, it has been noticed that objective assessment results fail to correlate properly with the reality most of the time. Considering this fact, many researchers prefer subjective assessment for inferring QoE of a video over the objective assessment method.

Table 1. MOS rating for subjective quality measurements.

MOS	Quality	Perception
5	Excellent	Imperceptible
4	Good	Perceptible
3	Fair	Slightly annoying
2	Bad	Annoying
1	Poor	Very annoying

2.2   Importance of Video Streaming

Video streaming is an attractive solution due to two important reasons. First, a client can start watching video immediately after selecting play option while video is being streamed instead of waiting for the entire video to be downloaded. In case when the video file size is big, assume a two-hour lecture video, it would take intolerably long time to download the entire file over a low speed Internet connection [6]. Second, video streaming would not consume much memory space in a user device. In order facilitate video streaming over Internet, a video streaming server first needs to encode the video file to reduce file size. The growing importance of video streaming service propelled significant research on designing video delivery protocols (e.g. real-time transport protocol, real-time transport control protocol) over Internet and video encoding mechanism (e.g. H.264) over the last couple of decades.

3   Proposed Class Lecture on Demand (CLD) Service

Our proposed approach composed of three steps: *(i)* understanding background traffic profile, *(ii)* evaluating MOS under peak and off-peak hours background traffic load profile for different number of video class-lecture clients, and *(iii)* providing QoE aware client admission policies. In order to test QoE of videos in a bandwidth constrained campus scenario, the overall experimentation procedures are stated in Fig. 1.

3.1   Understanding Background Traffic Profile

Campus network will be serving different kinds of traffic along with the video class-lecture clients. Therefore, we need to understand the behavior of traffic that the campus network serves (excluding video class-lecture traffic). We refer to this kind of traffic as background traffic in this paper. We produce a background traffic graph to investigate the traffic pattern in Fig. 2. Let $S = \{s_1, s_2, ..., s_N\}$ be the set of traffic load sample points from the traffic load profile depicted in Fig. 2.

3.2   Mean Opinion Score (MOS) Analysis

Based on the traffic profile depicted in Fig. 2, we are interested to find the maximum number of class-lecture video clients can be admitted at a given time (t). Let $M^{(t)}$ denote the number of class-lecture video clients and $B^{(t)}$ represents bandwidth consumed due to background traffic at time t. Assume, γ is the average bandwidth consumption of a class-lecture video at a certain compression rate (α) and BW_{max} is the maximum throughput of the campus network can offer. Then, the following condition should hold in order not to deteriorate QoE of the background flows.

$$\gamma M^{(t)} + B^{(t)} \leq BW_{\max}. \tag{1}$$

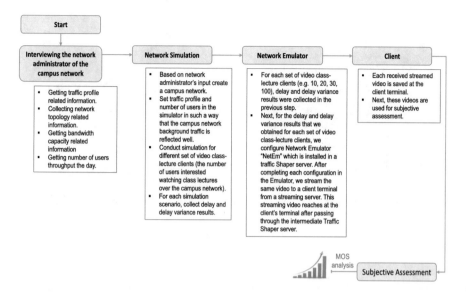

Fig. 1. Overall experimental procedures.

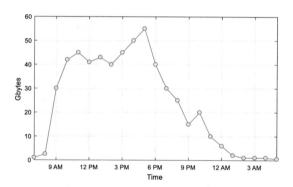

Fig. 2. 24 h traffic profile in a campus.

At this stage of our experiment, for different $B^{(t)}$ values, we check the MOS of different $M^{(t)}$ values for a certain compression rate of a video. Note that, in our experiment, we will be choosing two different kind of videos: fast video (consider the case in which lecturer has relatively high movement during class lecture period) and slow video (consider the case in which the lecturer has a few movement while delivering lecture at a class).

In our study, following the traffic profile illustrated in Fig. 2, we select upper boundary ($B^{upper} = max\{S\}$) and lower boundary ($B^{lower} = min\{S\}$) of the background traffic load. We consider that these two boundaries represent maximum and minimum traffic load during peak and off-peak hours of a campus traffic demand, respectively. We are interested to find at these two boundaries

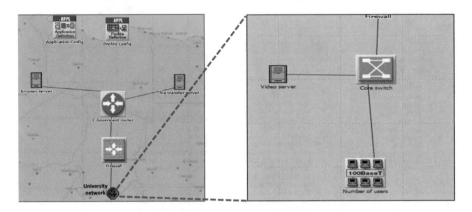

Fig. 3. Understanding campus network traffic performance during peak and off-peak hours using OPNET.

the maximum number of class-lecture video clients (M) can be accommodated. In order to do so, in OPNET simulator, we created background traffic for both B^{upper} and B^{lower}. We assume the background traffic is mainly FTP and HTTP traffic (Fig. 3).

We roughly estimated that 600 (B^{upper}) and 50 (B^{lower}) FTP and HTTP browsing background traffic flows in our simulation setup (in OPNET) would represent peak hours and off-peak hours traffic load presented in Fig. 2, respectively. We have several simulation scenarios in our experiment. For each scenario we configure in OPNET simulator, we collect delay and delay variance results. Here, below we explain the simulation setup and procedures:

> Step 1: For the slow video, we set B^{upper} background traffic and we create six simulation scenarios for different values of M (6, 36, 54, 66, 69 and 72).
> Step 2: We follow the same procedures stated in Step 1 for the fast video.
> Step 3: For each slow video, we set B^{lower} background traffic and we create six simulation scenarios for different values of M (6, 36, 54, 66, 69 and 72).
> Step 4: We follow the same procedures stated in Step 3 for each fast video.

Results (delay and delay variance results) from aforementioned simulation scenarios are recorded and tabulated in Table 2. From this table, we can infer that, for example, in case of slow video, during peak hours $(B^{upper} = 600$ FTP+HTTP traffic), if there are six video class-lecture clients (i.e. $M = 6$), the class-lecture videos would experience 1.6068 ms delay and 0.000008095 ms delay variance. Similarly, during the peak hours, if $M = 72$, video class-lecture clients would experience 924.05 ms delay and 135.1 ms delay variance. Upon completion of the network simulation, NetEm emulator is used in order to see how each of the delay and delay variance pairs presented in Table 2 affects a video. The setup is illustrated in Fig. 4.

Note that, we have 24 pairs of delay and delay variance results, as we can notice from Table 2. The NetEm emulator is configured to see the impact of each

Table 2. Delay and delay variance results for different M values during peak and off-peak hours of a day in a bandwidth constrained campus network.

Video type	Scenario	Number of users		Delay (ms)	Delay variance (ms)
		ftp + browsing	video		
Slow video (news)	Peak hours	600	6	1.6068	0.000008095
			36	1.9645	0.00007371
			54	2.84	0.00000006285
			66	4.1855	0.0003569
			69	4.524268	0.0011055
			72	924.05	135.1
	Off-peak hours	50	6	1.5705	0.000000002275
			36	1.8315	0.00001073
			54	2.5785	0.00017055
			66	5.0495	0.00030295
			69	6.09149	0.002585
			72	1693.8175	467.0275
Fast video (pamphlet man)	Peak hours	600	6	1.6068	0.000008095
			36	1.9645	0.00007371
			54	2.84	0.00000006285
			66	4.1855	0.0003569
			69	4.524268	0.0011055
			72	924.05	135.1
	Off-peak hours	50	6	1.5705	0.000000002275
			36	1.8315	0.00001073
			54	2.5785	0.00017055
			66	5.0495	0.00030295
			69	6.09149	0.002585
			72	1693.8175	467.0275

Fig. 4. Network emulator test-bed setup.

pair of delay and delay variance (the procedure is repeated for all the cases in Table 2 for both slow and fast videos). For each setup, video from the streaming server is streamed and saved at the client side after the video is shaped with the traffic shaper (see Fig. 4). Therefore, finally, we have 24 videos saved in client side, each of which represents the effect on video quality for each pair of delay and delay variance results stated in Table 2. The screenshots of some of the videos that are saved in client side are depicted in Figs. 5 and 6 for both peak hours (B^{upper}) and off-peak hours traffic load (B^{lower}) case. Looking at these figures, we can easily perceive that with the increment of video class-lecture clients (M) the video quality deteriorate (when $M = 72$ the videos are almost unrecognizable).

124   S. A. binti Haji Jalil et al.

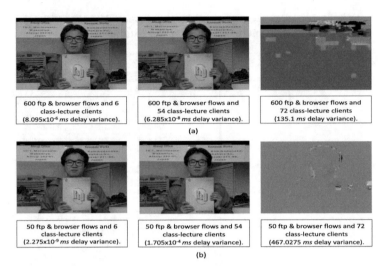

Fig. 5. Fast video quality for peak hours and off-peak hours background traffic under different M values: (a) peak hours scenario; and (b) off-peak hours scenario.

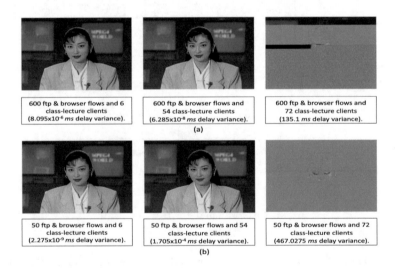

Fig. 6. Slow video quality for peak hours and off-peak hours background traffic under different M values: (a) peak hours scenario; (b) off-peak hours scenario.

In this paper, we use subjective assessment as it relates to our experiment. To get MOS, a subjective assessment with 20 human subjects are conducted. Out of 20 human subjects, there are 10 males and 10 females with age group of 20 to 33 years old. Each user (subject) is shown 24 video sequences of which 12 of them are slow movement videos called *News*, and the remaining 12 videos are fast movement videos called *Pamphlet Man*. For displaying the streaming video,

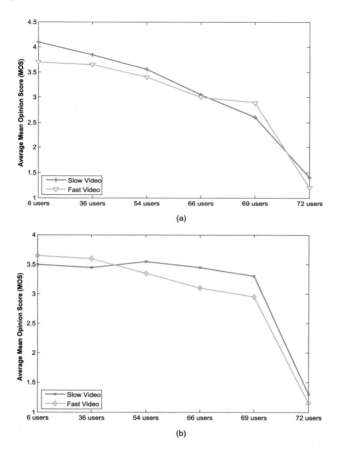

Fig. 7. Aggregated average MOS of videos during peak hours and off-peak hours under different value of M: (a) peak hours MOS (scenario considers B^{upper}); (b) off-peak hours MOS (scenario considers B^{lower}).

each user is allocated a laptop and the content is viewed only by a single user (discussion among the subjects is not allowed during this assessment).

Once the subjects watch all the videos without knowing the delay and delay variance have been added, they rate the videos based on the video quality they perceived. The video files are given different names, such as *Video1* to *Video24*. We do not want the subjects to have any clue about the delay and delay variance each of the videos experienced. In order to accomplish this, the videos are not played in descending or ascending order of delay and delay variance. After completing watching each of the videos, the subject rate his/her perception on the quality of the video (subject selects one of the five ratings on a continuous scale from 1 to 5, as shown in Table 1).

All feedbacks were recorded and gathered to calculate the average MOS per scenario. In Fig. 7, the aggregated average MOS for both slow and fast videos during peak hours (600 FTP and HTTP browser) and off-peak hours (50 FTP and HTTP browser) are plotted together as to make comparison of the average MOS. The graph shows that the average MOS decreases as M increases.

The lowest average MOS rating noticed in case of peak hours (see Fig. 7(a)). It shows that when $M = 72$ the average MOS rating of 1.2, indicating poor video quality. Figure 7(a) shows when $M > 66$ during peak hours viewers' rating for fast video is less than 3. This implies that perception based on video quality is annoying or very annoying. However, in case of off-peak hours (see Fig. 7(b)), when $M > 69$, viewers have the same opinion. Therefore, we surmise, obviously, we can accommodate more video class-lecture clients (M) during off-peak hours than the peak hours.

By comparing the peak and off-peak hours MOS results, the QoE of both slow and fast video abruptly decreases as the delay variance and delay increase (due to increment of the value of M). Both peak hours and off-peak hours ratings are approximately similar for both slow and fast videos. Despite that, most of the fast videos show a lower average MOS compared to slow videos. A fast video is more affected with delay variance because it has a more movement compare to a slow video.

3.3   QoE Aware Video Class-lecture Admission Policies

It has been noticed that with the increased number of M (video class-lecture clients), the video quality deteriorates as noticed in Figs. 5, 6 and 7. When video streaming service is in operation, the value of M should be set carefully during peaks and off-peak hours of a day. For instance, if the steaming server wants to meet MOS greater than or equal 3, the number of CLD clients should not be more than 66 and 69 during peak hours and off-peak hours, respectively, as observed from Fig. 7. Overall, taking the background traffic load into account, the streaming server should be configured in such a way that MOS requirement threshold does not get violated for the video class-lecture clients.

Our study also reveals that the lectures' movement behavior could influence MOS. A lecturer with high movement relatively during the class lecture capturing would be more affected with packet delay variance. In turn, this would result in reducing MOS compared to the case when the lecturer has very slow movement during capturing a lecture (slow video).

4   Recommendations

Based on our study, we have the following suggestions that would improve class-lecture video streaming performance at a bandwidth constrained campus network.

– As an alternative to video streaming in a bandwidth-constrained campus during peak hours, students can be encouraged to download the class lectures during off-peak hours. It would be more convenient if one student downloads and share them with the whole class.

- Video segmentation is a popular approach [9]. The video can be segmented based on some criteria (e.g. topic, class intervals, chapter). Using this approach, instead of downloading the entire class lecture, a student may download the portion that the student wants to watch only, thereby reducing bandwidth consumption and streaming server load.
- Dynamic video quality selection considering the traffic load at a campus at a given time would be useful similar to different existing video streaming services (e.g. YouTube).
- In case when a particular video is highly requested from the students, multicasting mechanism can be applied [11,12] to reduce bandwidth wastage.
- If many students are interested on a particular video lecture at a given time, the video scheduler can set higher resolution of the video for that class lecture. This will improve the MOS for that video lecture and in return contribute in increasing overall users' satisfaction (QoE).
- It is recommended for the campus network administrator to increase the number of servers and place them at strategic locations in order to reduce network bottleneck points. For example, one way to reduce the load from the centralized server is to place the streaming server close to a dormitory if many students are residing there.
- Peer-to-peer video streaming is a popular mean to reduce load from the streaming servers of video service providers [10]. Aside from the campus owned servers, devices of students—if they are interested—can be used for class lecture storage and streaming purposes.
- Lecturers need to put effort in order to deliver lectures efficiently (make the class-lecture videos concise and effective).
- Lecture capturing process needs to be efficient [2]. Additionally, during postprocessing of a captured class-lecture video, no class activity periods can be removed (if there is any) in order to reduce video size of the lecture.

5   Conclusion

In this paper, we proposed a background traffic load aware class-lecture video streaming solution for a bandwidth constrained campus network. Aside from this, for such network, we provided several recommendations to make class-lecture video streaming as a feasible approach. In order to infer QoE performance of class-lecture clients when a bandwidth constrained campus adopts CLD service, we provided a detailed experimental procedures in this paper. In the future research, we aim at incorporating: (i) a QoE and background traffic load aware optimal video quality finding mechanism and (ii) a knowledge (class lecture prerequisites) dependency based video lecture tagging solution.

Acknowledgment. This work was supported by grant No. UTB/GSR/1/2017 (9) from the internal project grant of Universiti Teknologi Brunei (UTB), Brunei Darussalam.

References

1. Bhamidipati, V., Kilari, S.: Effect of Delay/Delay Variable on QoE in Video Streaming. M.S. thesis, School of Computing at Blekinge Institute of Technology, Karlskrona, Sweden, May (2010)
2. O'Callaghan, F.V., Neumann, D.L., Jones, L., Creed, P.A.: The use of lecture recordings in higher education: a review of institutional, student, and lecturer issues. Educ. Inf. Technol., 1–17 (2015)
3. Hecht, J.B., Klass, P.H.: The evolution of qualitative and quantitative research classes when delivered via distance education. Paper presented at the Annual Meeting of the American Educational Research Association, Ontario, Canada, (ERIC Document Reproduction Service No. ED 430 480)
4. Kim, H.J., Choi, S.G.: A study on a QoS/QoE correlation model for QoE evaluation on IPTV service. In: Proceedings of 12th International Conference on Advanced Communication Technology (ICACT), vol. 2, pp. 1377–1382 (2010)
5. Mwela, J.S., Adebomi, O.E.: Impact of packet losses on the quality of video streaming. M. Sc. Thesis School of Sience in Electrical Engineering School of Computing at Blekinge Institute of Technology, Karlskrona, Sweden (2010)
6. McCrohon, M., Lo, V., Dang, J., Johnston, C.: Video streaming of lectures via the internet an experience. In: ASCILITE 2001 - The 18th Annual Conference of the Australasian Society for Computers in Learning in Tertiary Education, Melbourne, pp. 397–405 (2001)
7. Mok, R.K.P., Chan, E.W.W., Chang, R.K.C.: Measuring the quality of experience of HTTP video streaming. In: 12th IFIP/IEEE International Symposium on Integrated Network Management (IM 2011) and Workshops, Dublin, pp. 485–492 (2011)
8. Wang, Z., Li, L., Wang, W., Wan, Z., Fang, Y., Cai, C.: A study on QoS/QoE correlation model in wireless-network. In: Asia-Pacific Signal and Information Processing Association Annual Summit and Conference (APSIPA), pp. 1–6 (2014)
9. Chien, S.-Y., Huang, Y.-W., Hsieh, B.-Y., Ma, S.-Y., Chen, L.-G.: Fast video segmentation algorithm with shadow cancellation, global motion compensation, and adaptive threshold techniques. IEEE Trans. Multimed. **6**(5), 732–748 (2004)
10. Dhungel, P., Hei, X., Ross, K.W., Saxena, N.: The pollution attack in P2P live video streaming: measurement results and defenses. In: P2P-TV 2007 Proceedings of the 2007 Workshop on Peer-to-Peer Streaming and IP-TV, pp. 323–328 (2007)
11. Newaz, S.H.S., Bae, Y., Lee, J., Lee, G.M., Crespi, N., Choi, J.K.: Minimizing the number of IGMP report messages for receiver-driven layered video multicasting. In: IEEE VTC Spring, May 2011
12. Holbrook, H.W., Cheriton, D.R.: IP multicast channels: EXPRESS support for large-scale single-source applications. In: Proceedings of the Conference on Applications, Technologies, Architectures, and Protocols for Computer Communication (SIGCOMM 1999). ACM (1999)

Green Computing

Big Data Analytics and IoT Gadgets
for Tech Savvy Cities

N. Venkata Ramana[1], Puvvada Nagesh[1], Soni Lanka[2,3,P], and Rama Rao Karri[4(✉)]

[1] Koneru Lakshmaiah Education Foundation, Department of CSE, Vaddeswaram,
Vijayawada, AP, India
ramana.9n@gmail.com, pnagesh.qa@gmail.com
[2] Department of Computer Sciences and Systems Engineering, Andhra University,
Visakhapatnam, India
karri.sony@gmail.com
[3] Computer Sciences, Faculty of Science, Universiti Brunei Darussalam,
Gadong, Brunei Darussalam
[4] Faculty of Engineering, Universiti Teknologi Brunei, Gadong, Brunei Darussalam
karri.rao@utb.edu.bn

Abstract. In towns and urban areas to make it as a smart and to address the issues of urban open and the city advancement shrewdly, the utilization of IoT gadgets, and the savvy framework is the quick and profitable source. In any case, interconnecting a large number of IoT gadgets while speaking with each other over the Web to build up a keen frame work, brings about the age of colossal measure of information, named as huge data or a big data. To incorporate IoT benefits so as to get constant city information and afterward preparing such enormous measure of information in a productive route went for setting up brilliant city is a testing assignment. In this manner, we processed and built up a smart city framework in light of IoT utilizing big data and Investigation utilizing Flume and Hive. Hive is a rank sharing center foundation device to process ordered data in Hadoop. It dwells over Hadoop to abridge Huge Information. It can be utilized for dumping movement information in Hadoop Circulated Record Framework. We utilize sensors structure including keen home sensors, vehicular systems administration, climate and water sensors, smart stopping sensors, observation objects, and so on. The total engineering and usage is proposed, which is executed utilizing Hadoop biological community in a genuine domain. The framework usage comprises of different advances that begin from information age and gathering, collecting, filtration, characterization, preprocessing, figuring and completed at basic leadership. The framework is for all purposes actualized by taking city information source to create a smart city. The work establishes the anticipated framework is effective and versatile environment.

Keywords: Big data · Flume · Hadoop · Hive · IoT

© Springer Nature Switzerland AG 2019
S. Omar et al. (Eds.): CIIS 2018, AISC 888, pp. 131–144, 2019.
https://doi.org/10.1007/978-3-030-03302-6_12

1 Introduction

The pattern of living is presently changing to smart way. One of the report advised that in 2050, 70% of the total populace will live in city's [1]. Thus, a fast increment has been found in the progress of the populace towards urban areas. Along these lines, for urbanization, it is a most outrageous basic feature to understand the enthusiasm for advantage profile to improve the capability and may gain the continuous movement the town association. Straightforwardly, couple of affiliations are going with on IoT stages in live watching, masterminding and collecting urban process parameters. For Example Japan's broadband access is giving the workplace of correspondence between people, people and things, and things and things [2]. So, Korea's savvy home empowers their kin to get to things remotely [3]. Mumbai cutting edge I-Hub [4] aims to grasp the cutting edge "U" type organize through a protected and pervasive system [5]. Consequently, the utilization of IoT for keen frameworks brings about improving the quantity of things of objects to be interconnected with each other, which brings about the mind-boggling measure of the mixed information, alluded to enormous information. Separating such type of data in light of the customer needs and choices, the urban zones would end up being considerably shrewder. In any case, the examined framework works at a restricted level, without thinking about the significance of Huge Information age and taking care of. Such exercises ought to be trailed by the measure of information gathered, disconnected and continuous big data handling [6–8] and examination, and basic leadership. More often than not, information gathering and examination methods are hard to accomplish in such condition. Hence, there is a need to consolidate brilliant innovation that could effectively gather the information, performed examination, take constant choices and foresee the prospect for better city arranging and improvement.

Have comprehended the practicality and capability of the IoT and the new frameworks, So in this article, we push the idea of frameworks toward the smart city improvement in view of big data investigation. In this work, we planned the entire design to the created smart brilliant city and did urban arranging utilizing IoT and big data analytics. The framework gives the direction to the private sector and administration to make their urban communities more quick witted and keen with a specific end goal to take a choice at continuous in view of current city situations. The entire framework portrayal is delineated in next categories.

2 Motivation

City development and planning of savvy city apps have major crash on the life style of public [1]. This incorporates the impact of resident as far as wellbeing and safety, disaster management, contamination control, et cetera so forward. Diverse tasks identified with checking of cyclist, autos, car parking, and so on are experiencing that uses sensors administrations for the accumulation of particular gathering of information. Obviously, extraordinary other administration area applications are distinguished that uses shrewd city IoT foundation to arrangement tasks in air, commotion, contamination, and observation framework in the urban communities.

Nonetheless, in the advancement of any city, the vehicle framework has a key part. Indeed, even a nation can just advance quickly if his shipping framework and the offices of shipping to the resident are momentous. A decent shipping framework makes the errand to be performed instantly. The more astute shipping has a considerable measure of different advantages, for example, diminishment of contamination, help of natives, fast advancement, economy change, and numerous more. The ongoing exploration comprises of a not very many research discoveries in the field of brilliant movement and also in the keen city. In addition, in this electronic period, where billions of gadgets are associated on the web create a large number of terabytes of Enormous Information. The huge information produced by the different IoT frameworks is utilized to investigate diverse parts of smart city. A comparative idea is taken after utilizing the IoT worldview and the huge information ideas for urban arranging. In this manner, to break down such measure of information and settle on shrewd choices is a noteworthy test. In this manner, in light of the need of the nationals and experts, we center around the working of city the all the more shrewdly by giving continuous data with respect to the city for the residents also specialists.

Hence, the justification behind our expectations is to enhance the immense arrangement of ICT assets in building up the whole framework. At this time we come to realize that the succession of late modernization in the implanted framework delineates the patterns of ICT. In this way, a framework breathe in the greater part of the ongoing advancements in the field of shrewd urban communities, because of which an astounding development can be found in a not so distant future. The outline of this framework requires every one of the abilities of detecting the earth and examining the detecting data. In this manner, different continuous activity could be invited because of these innovative assets. In addition, it can be seen that incorporating a lot of information to play out the effective investigation are as of now performed taking care of business. In any case, with extensive scale condition, it is inevitable that the colossal part of information is left put out of articulation. Therefore, such information can't give us a superior comprehension of the circumstance so we may anticipate prospect.

3   Background and Associated Effort

Huge information and its assessment are at the skirt of present day knowledge and production, where creator features the personality of number of sources on huge information, for example, online exchanges, emails, audios, recordings, search queries, well-being records, social network communications, pictures, click-streams, logs, posts, look questions, wellbeing records, person to person communication co-operations, cell phones gadgets and applications, logical gear, and signals [9]. The planned display is utilizing regular data storage devices. The experiment is to seize, shape, store, oversee, share, dissect, and imagine the big data. Similar investigations have been performed on savvy urban communities working limitedly [10] and factual highlights of the big data also [9–14]. In a good number of the cases, we don't depend on their realistic estimations. With an observation to confirm the realistic estimations of the enormous informational indexes, there is a chance to talk to them in a few different structures, i.e., how to portray huge informational collections as most of the information preparing errands depend on some fitting information portrayal. For example, in the event of picture preparing situation, the wavelet change [15, 16], if there should arise an occurrence of remote detecting situation, multi-determination portrayal, for example, picture division [17], picture deionizing [18], picture reclamation [19], picture combination [17], change location [18], include extraction [19], and picture elucidation. Thus, evaluation of factual estimations of big data in the wavelet change space continuously or disconnected is a key testing zone.

Flume: It is an appropriated, strong, and available framework for capably assembling, conglomerating, and moving a great deal of log data. It has a clear and versatile outline in light of spouting data streams. It is fiery and accuse tolerant with tunable relentless quality frameworks and various failover and recovery instruments. It uses a direct extensible data demonstrate that thinks about online indicative application.

The reputation big data is utilized for change and administration of urban areas with the assistance of information surge that give significantly more modern, more extensive scale comprehension and control of refinement. To the best of our imminent, there is no such modern meaning of enormous information, yet a review of developing patterns connote various key highlights, for example, volume demonstrates a great deal of information to be amassed for handling and investigating while speed alludes to the fast preparing and examination (e.g., web based spilling, ongoing remote detecting, social sites information, e-wellbeing information, et cetera.). Then again, the term Assortment alludes to the unfathomably changed formats (e.g., one Machine-to-another Machine, Web of belongings, Remote antenna System, et cetera).

If there should arise an occurrence of coordinated effort collaborations inside logical groups through various learning fields, brings about new items. Since the part of groups and systems as a fundamental state of oddities is examined in [1]. IoT is relied upon to be a considerable help for the Information and technology framework of shrewd urban areas [20, 21]. A plan is additionally displayed that sends signals to a cloud that give IoT administrations are introduced in [22]. This effort examined diverse business cloud-helped remote detecting stages and features their capacities. The idea of signals is to produce a huge size of huge information.

HIVE: Hive is a foundation tool in information distribution center to process arranged or organized information in Hadoop structure. It is over the Hadoop structure to preparing designed information, and makes questioning and dissecting simple. Apache Hive (HiveQL) with Hadoop Distributed document System is utilized for Analysis of information. Hive gives a SQL-like interface. To set up HIVE in Hadoop In this manner, the appropriation of setting to signal information for starting understandable and convincing data assumes a basic part. Additionally, setting mindful registering has turned out to be effective in understanding tangible information. Different works about setting a mindful in terms of IoT perspective is examined in [23]. Because of the way that the sending of a most extreme number of signals alongside various heterogeneous gadgets and frameworks, and because of untrustworthy nature of the greater part of articles, the quality and services are to be decreased. To take care of such issue, the intellectual administration structure for IoT is anticipated [23]. This plan, insightful and self-governing execution of various apps are empowered by putting perception and nearness with related articles for the appliance. As a result of the way that urban master-minding and change apps can be benefitted from a savvy city. IoT capacities can be gathered into effect territories [1]. That incorporates the impact on the resident as far as wellbeing and security, the transportation framework regarding portability and contam-ination, et cetera so forward. Diverse activities identified with checking of cyclist, autos, open auto stopping, and so forth are experiencing that uses sensors administrations for accumulation of particular gathering of information. Clearly, extraordinary other admin-istration space applications are recognized that uses keen city IoT foundation to arrange-ment tasks in air, commotion, contamination, vehicle portability, and reconnaissance framework in the urban communities. The ongoing examination comprises that not many more research discoveries in the field of brilliant city and additionally in municipal regions. So also, a minimized framework isn't yet manufactured which is more adaptable and productive. The huge information is utilized to break down various parts of the keen city and after that uses the learning get from the past created information for the advancement of urban areas. A comparable idea is taken after utilizing the IoT world-view and the big data ideas for urban arranging.

Along these lines, we endeavored to think of an answer that is material in the shrewd city and additionally in the city regions. The planned framework is actualized and tried on the Hadoop structure with a start to get the ongoing impacts on account of constant brilliant city choice. Also, Hadoop frame work and Map Reduce technique is utilized for extensive verifiable information for municipal arranging and future upgrades. Utilizes the learning get from the past produced information for the improvement of urban areas. A comparable idea is taken after utilizing the IoT worldview and the huge information ideas for urban arranging.

In this manner, we endeavored to think of an answer that is material in the shrewd city and in addition in the city territories. The planned framework is actualized and tried on the Hadoop framework with a start to get the continuous impacts on account of constant keen city choice. In addition, Hadoop frame work and Map Reduce technique is utilized for vast chronicled information for urban arranging and future upgrades.

4 Convenience IoT-Based Savvy City

4.1 Structure Overview

The essential idea of the savvy city is to collect the correct data at the ideal point on the correct gadget at the perfect time to settle on the city associated choice with ease and to encourage the national's all the more snappy and quick ways. To build up the IoT-based savvy city, we sent a few remote and wired signals, reconnaissance lance cameras, crisis catches in venues, and other settled gadgets. The primary test in such manner is to accomplish keen city framework and connection savvy framework produced informa-tion at one place. We do this by setting the principle information center point connecting all keen frameworks to have all savvy framework information at a one place. Figure 1 demonstrates the signals and smart framework organization keeping in mind the end goal to create information utilizing a focal center point for building the bright city. With a specific end goal to get Continuous city information, we projected to convey numerous signals at better places to accomplish savvy homes, parking, whether and wet resources frameworks, vehicle movement, condition populace and observation framework. These frameworks are utilized by the experts to settle on an astute choice in light of the ongoing information to set up the savvy town.

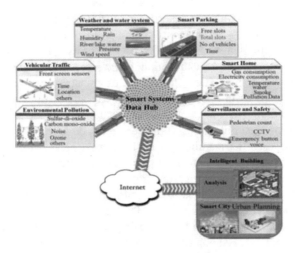

Fig. 1. Light sensors operation.

This savvy house, the house is constantly checked by feeding information produced from the signals, e.g., the burn and heat to recognize a flames at continuous, the power and swap gossip utilization to successfully deal with the influence, swap gossip, and hose utilization to the homes and distinctive zones of the city. The savvy leaving helps in dealing with the carriers going back and forth out of various leaving zones. Along these lines by savvy leaving analysis, the requirement for new leaving at particular zones can be distinguished. In our framework, the nationals effortlessly get the data of the closest free leaving slot and reasonable placing for stopping. Climate and water framework gives the climate related information like temperature, moistness, rain, water levels and wind speed at waterways, lakes, dams, and different stores. On the planet, the vast majority of the surge happens because of the rain and comparatively few by flurry liquefying and block crack.

Along these lines, we utilize rain estimating signals and snow softening constraints keeping in mind the end goal to foresee the surge prior. We likewise foresee about the water supplies ahead of time with a specific end goal to address the issue of the drinking water to natives. Carrier's activity data is the most critical wellspring of a shrewd town. Through this sort of information source and with helpful continuous examination the resident and also administration can get more advantages. In this smart city framework, we are accessing the activity data by GlobalPRS, carrier's signals, and also the signals set on the face computer screen of the auto. We get the area of every carrier the quantity of carrier between two sets of signals put at the different area of the city.

In addition, if any mishap happens, the face monitor will be harm and the signals will send the alarm to the police force, movement experts, and healing center. In addition, a town can never be smart with unfortunate nationals. Along these lines, while outlining shrewd town, we put a different component to get condition information this incorporates gases data, for example, specific metals1, carbon1 monoxide1 sulfur dioxide1, ozone1, and commotion and additionally. The people are alert when any of the toxic1 ga1s is extra in the atmosphere.

To wrap things up, the most critical object for the general population of the well-groomed town is the protection concerns. Protection is accomplished by the planned framework by persistent observing the record of the entire town. In any case, this is exceptionally hard to break down all town recordings and distinguish any accident with anybody at continuous framework. To conquer this restriction, we recommend new situations, which increment the safety of the arrangement of the entire city. We put different crisis catches including receivers at better places of the city with observations cameras. At the point when any incident occurs with anybody like theft, auto stolen, and so forth. He can simply push the crisis catch at any close police headquarters and so on. Along these lines, the police or security organizations can begin observing the adjacent areas through reconnaissance cameras and can rapidly find the faker.

The execution reproduction base on the necessities of the brilliant town, we anticipated an engineering and the usage to break down IoT-based savvy frameworks to build up keen city by constant analysis. The framework finish design and actualized models is given in Fig. 2. It demonstrates the full points of interest of the considerable number of steps performed from information age to basic leadership and applications. At first, every framework will make their information, for example, brilliant homemade information, carrier information, savvy stopping information, and so forth. At each framework, there is hand-off hub, gathering information from all signals and after that sending to the web and the focal information Hub. Astute City Building is the fundamental examination framework that is in charge of all exercises from data aggregation, filtration to Decision making. It contains particular servers prepared with different progressed illustrative calculations at various levels. At information pre-processing stage, the information is accumulated by snappy framework cards and driver, with the objective that it can't lose any packs. As the sensors have a considerable measure of metadata, and the signals in addition make the mixed kind of information.

Subsequently, all the inconsequential data about data and the abundance data are not needed at filtering system. The order system arranges the approaching information from the different framework by the significance. After request, the portrayed data is changed over to the created shape, i.e., justifiable to the Hadoop environment, for example, arrangement document or unthinkable for with parameters data.

Since we are taking care of with a gigantic measure of information (named as large information). In this way, we require a framework that could proficiently process an enormous size of continuous (savvy city) and in addition disconnected (urban arranging) information. To reach these prerequisites, we utilized Hadoop biological system, which contain central hubs and hive and flume apparatuses, and different information hubs under the central hub. The Hadoop biological system has HDFS record stockpiling with

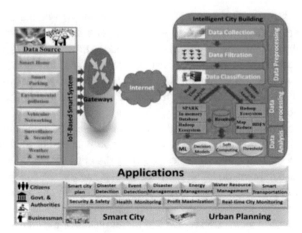

Fig. 2. System construction and derived model.

various areas of information, which isolates the information into an equivalent measure of parts and put away them in numerous information hubs.

Afterward, the parallel preparing is performed on these parts utilizing MapReduce framework. Hadoop fundamentally utilized for bunch handling, however, so as to utilize it for constant investigation, we utilized flume over the Hadoop framework. All the preparing counts, comes about age are done at Hadoop biological community with help of hive and flume.

At long last, the basic leadership is performed in light of the outcomes produced by Hadoop biological community. The basic leadership approach utilizes machine learning, design acknowledgment, delicate processing and choice models. The created comes about are utilized for some, brilliant city actuates, for example, appeared as apps in shape 2.

4.2   Framework Completion and Data Gathering

We execute the entire framework on Hadoop biological system taking it as the Clever City Building. All the savvy frameworks are associated with the primary framework, which gathers the ongoing information. All the rest of the exercises are done by the hive distributed biological system. We built up an equipment based on vehicular system and a support station with three vehicles, (see Fig. 3) to produce genuine information, for example, area, speed, time, and so on. The support station is joined to the framework by using USB port that forwards the vehicular data to a framework. For all the other brilliant frameworks utilized accessible savvy frameworks data sets. The constant activity is handled by Hadooppcap-lib, Hadoop-pcap-scr-de libraries and changed over into grouping record to make it equipped for preparing on Hadoop.

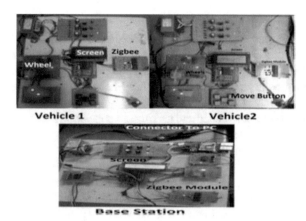

Fig. 3. The implemented vehicle network.

Every office of the keen town is executed as an isolated function. Natives have restricted access to the consequences of these functions, and the legislature has occupied. As it is extremely mind boggling right now to execute the keen frameworks, consequently we take existing datasets of the specified brilliant frameworks from different solid assets. The points of interest of each data set including the data sets estimate, the zone data, depiction, the quantity of the references and parameters are shown in Table 1.

Table 1. Datagroup details

Sl. no	Dataset	Area	Description	No of attributes	Ref.
1	Water use	Guntur	1000 House meter reading	11	29
2	National high way traffic	IBM	Location, speed	5	30
3	Vehicular	Jupudi	24 h, 1200 cars	5	31
4	Parking	Guntur city	8 Parking	7	32
5	pollution	IBM	449 sensor ozone	8	32
6	City traffic	Eluru	Sensors b/w two points	9	32
7	Weather	Karempudi	Temperature, humidity, rain, pressure, air flow speed	7	32

5   Information Analysis

The information investigation is done by utilizing the created framework. The investigation results are utilized for some reasons and profited the general public from normal native to agent and the administration experts. Here, we are featuring a portion of the effects and advantages picks up by nationals and administration by genuine information investigation. Increasing Relevant Use of the Resources with Internet to Tell Us when and where to Save It's estimated that major cities waste up to 50% of water due to leaky pipes. Irrigation systems run when it's raining and street lights remain on even when there's no activity in the area. Internet connected sensors can control detect unnecessary use and make adjustments. Irrigation systems can turn off when rain is detected, lights can go dim when they aren't needed and leaking pipes can send a text to landlords. These improvements help us save resources by increasing efficiency and large amounts of money. The Internet of Things can be used to empower the systems we've already created to work for us in the best way possible. For instance, public transportation could work even better with IoT and cloud based big data automation enabling operators and control center to see their position in route and ticket sales. With large displays and through mobile interfaces Bus riders could see real time positions and get notifications in time for the bus. It avoid time wastage at the stop or missing the bus. Revenue could be generated through location-based advertising.

For example, citizens can monitor the pollution concentration in each street of the city or they can get automatic alarms when the radiation level rises a certain level. In case of emergency management smart metering to monitor the optimum usage of energy, gas water etc. Figure 4(a) charts created to demonstrate the normal vehicle's speed about when the force of the vehicles is low and high individually. Also, Fig. 4(b) demonstrates the power of the activity among three focuses at a different time. Figure 5(a) is the diagram created by the framework to demonstrate the evaluated time to reach to the next point in light of the normal speed of the vehicles1 and the power of the vehicles. The framework likewise looks at the over speed vehicle's. The over speed region of the Madrid Thruway regarding various vehicles are appeared in Fig. 5(b).

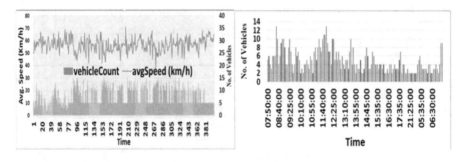

Fig. 4. (a) Velocity of cars and buses at less force of movement among two points (b) No. of cars and buses among two sources and goal at different instance

From this examination, A resident can choose which is the appropriate courses to achieve the goal relying upon the on the present movement situation. Also, the administration can control movement and make improve arrangement at runtime to diminish the force of the activity on the swarmed street. Open the infertile street due to any occurrence during run-time. Likewise, the administration can do numerous objects from current movement investigation.

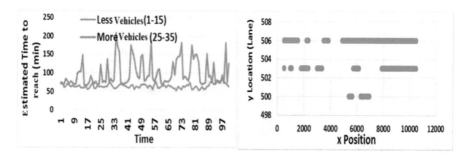

Fig. 5. (a) Predictable time to arrive at the target location (b) Area of speed contravention on National NH-5

The vehicles parking investigation in the city are appeared in Fig. 6(a) and (b) that depict the free and devoured spaces of a few parking areas on different time slots. From such kind of examination comes about, the nationals are refreshed to choose the best appropriate parking area close to their area. People can spare fuel without physically looking through the without charge car carport. More or less, it makes the benefit harmony between the low and high-benefit traders by redirecting client to free parking garages. Also, the legislature can influence urban wanting to assemble all the more stopping regions close to the spots where the greater part of the general population typically go.

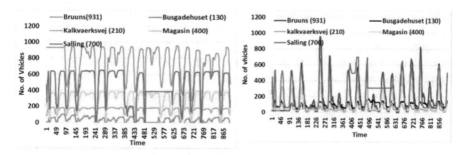

Fig. 6. (a) Free Spaces at different parking points at various circumstances (b) Parking lots usage

From shrewd home information, the legislature can give a great deal of administrations to the nationals and can likewise settle on savvy constant choices at run time, for example, fire discovery and control, vitality administration, and so on. Here we are simply taking one instance of water administration, which benefits the legislature with respect to legitimate control and arranging of water use. Figure 7(a) demonstrates the utilization of water for different computed by the framework from each home utilization. The created framework persistently screens the earth for poisonous gases. The checking diagram for a brief span, when a portion of the gases surpass, is appeared in Fig. 7(b). The framework produces alarms when one of the gas surpasses the unsafe edge restrain. It modifies the general population particularly the people with the specific sicknesses snot to go out. The administration can likewise do urban arranging, i.e., anticipate activity, town and mechanical construction and moving to different spots in light of the city contamination.

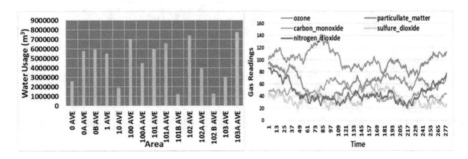

Fig. 7. (a) Usage of water in rural areas in Guntur. (b) Different categories of pollution in 24 h

6 System Evaluation

We executed the framework utilizing a few (4–5) Hadoop distributed information hubs. Be that as it may, for assessment reason, one system configuration on SOLARIS 14.04 LTS coreTMi7 machine with 4.2 GHz processor and 16 GB Ram. Since the framework depends on huge information examination, thusly, the frame works is assessed as for productivity and reaction time utilizing disconnected activity and also continuous movement. The preparing time comes about as for disconnected data sets.

Information size and handling moment are specifically relative to each other. Furthermore, throughput is additionally straightforwardly near to information estimate on account of the parallel handling environment of Hadoop framework, this is the significant accomplishment of the framework.

7 Summary and Future Scope

The Savvy city has a more impact on cities' financial system. Savvy city systems is a on the spot decision making system. This paper concentrates on the execution of the savvy city and savvy parking, savvy whether reports by the use of the Internet of things and Hadoop flame and hive frame works. Different clever frameworks are utilized to get constant city information to settle on a choice. The Hadoop, hive and flume biological community is utilized to analyze Big Data produced by savvy frameworks sent in the town. The structure is essentially actualized and tried on genuine information.

Further enhancement to the above work is arranging the genuine organization of every Smart framework, testing the precision of the framework, and we focus on considering reliability and accuracy.

References

1. Jin, J., Gubbi, J., Marusic, S., Palaniswami, M.: An information framework for creating a smart city through internet of things. IEEE Int. Things J. **1**, 112–121 (2014)
2. Ramana, N.V., Kumar, P.V.M.S., Nagesh, P.: Analytic architecture to overcome real time traffic control as an intelligent transportation system using big data. Int. J. Eng. Technol. **7**, 7–11 (2018)
3. Ramana, N.V., Nagesh, P., Kumar, P.V.M.S.: IoT based scientific design to conquer constant movement control as a canny transportation framework utilizing huge information available in cloud networks. J. Adv. Res. Dyn. Control. Syst. 1395–1402 (2018)
4. Vignesh, U., Sivakumar, Venkata Ramana, N.: Survey and implementation on classification algorithms with approach on the environment. Int. J. Eng. Technol. (IJET) **7**, 438–440 (2018)
5. Srivastava, L.: Japan's ubiquitous mobile information society. info **6**, 234–251 (2004)
6. Karri, R.R., Sahu, J.N.: Modeling and optimization by particle swarm embedded neural network for adsorption of zinc (II) by palm kernel shell based activated carbon from aqueous environment. J. Environ. Manage. **206**, 178–191 (2018)
7. Madhavi, R., Karri, R.R., Sankar, D.S., Nagesh, P., Lakshminarayana, V.: Nature inspired techniques to solve complex engineering problems. J. Ind. Pollut. Control. **33**, 1304–1311 (2017)
8. Karri, R.R., Chimmiri, V.: Nonlinear model based control of complex dynamic chemical systems. Adv. Chem. Eng. Res. (2013)
9. Giroux, S., Pigot, H.: From smart homes to smart care: Icost 2005. In: 3rd International Conference on Smart Homes and Health Telematics. IOS Press (2005)
10. Han, S.S.: Global city making in Singapore: a real estate perspective. Prog. Plan. **64**, 69–175 (2005)
11. O'Droma, M., Ganchev, I.: The creation of a ubiquitous consumer wireless world through strategic ITU-T standardization. IEEE Commun. Mag. **48**, 158–165 (2010)
12. Xia, F., Yang, L.T., Wang, L., Vinel, A.: Internet of Things. Int. J. Commun. Syst. **25**, 1101–1102 (2012)
13. Lu, J., Li, D.: Bias correction in a small sample from big data. IEEE Trans. Knowl. Data Eng. **25**, 2658–2663 (2013)
14. Soni, L., Madhavi, M.R., Abusahmin, B.S., Puvvada, N., Lakshminarayana, V.: Predictive data mining techniques for management of high dimensional big-data. J. Ind. Pollut. Control. **33**, 1430–1436 (2017)
15. Krishna, M., Chaitanya, D.K., Soni, L., Bandlamudi, S.B.P.R., Karri, R.R.: Independent and distributed access to encrypted cloud databases. In: Advances in Intelligent Systems and Computing (2018)
16. Li, R., Lin, D.K.J., Li, B.: Statistical inference in massive data sets. Appl. Stoch. Model. Bus. Ind. (2012)
17. Cormode, G., Garofalakis, M.: Histograms and wavelets on probabilistic data. IEEE Trans. Knowl. Data Eng. **22**, 1142–1157 (2010)
18. Yang, Q., Chen, Y., Xue, G.-R., Dai, W., Yu, Y.: Heterogeneous transfer learning for image clustering via the social Web. In: Proceedings of the Joint Conference of the 47th Annual Meeting of the ACL and the 4th International Joint Conference on Natural Language Processing of the AFNLP, ACL-IJCNLP 2009, vol. 1. Association for Computational Linguistics (2009)
19. Portilla, J., Simoncelli, E.P.: A parametric texture model based on joint statistics of complex wavelet coefficients. Int. J. Comput. Vis. **40**, 49–70 (2000)

20. Mallat, S.G.: A theory for multiresolution signal decomposition: the wavelet representation. IEEE Trans. Pattern Anal. Mach. Intell. **11**, 674–693 (1989)
21. Shah, V.P., Younan, N.H., Durbha, S.S., King, R.L.: Feature identification via a combined ICA–wavelet method for image information mining. IEEE Geosci. Remote Sens. Lett. **7**, 18–22 (2010)
22. Liu, P., Huang, F., Li, G., Liu, Z.: Remote-sensing image denoising using partial differential equations and auxiliary images as priors. IEEE Geosci. Remote Sens. Lett. **9**, 358–362 (2012)
23. Srikanth, S.V., Pramod, P.J., Dileep, K.P., Tapas, S., Patil, M.U., Sarat, C.B.N.: Design and implementation of a prototype smart parking (SPARK) system using wireless sensor networks. In: 2009 International Conference on Advanced Information Networking and Applications Workshops. IEEE (2009)

Improving Accuracy of Dissolved Oxygen Measurement in an Automatic Aerator-Control System for Shrimp Farming by Kalman Filtering

Jessada Karnjana[1], Thanika Duangtanoo[1(✉)], Seksun Sartsatit[1],
Sommai Chokrung[1], Anuchit Leelayuttho[1], Kasorn Galajit[1,2],
Asadang Tanatipuknon[2], and Pitisit Dillon[3]

[1] NECTEC, National Science and Technology Development Agency,
112 Thailand Science Park, Khlong Luang 12120, Pathum Thani, Thailand
{jessada.karnjana,thanika.duangtanoo,seksun.sartsatit,
sommai.chokrung,anuchit.leelayuttho,kasorn.galajit}@nectec.or.th
[2] Sirindhorn International Institute of Technology, Thammasat University,
131 Moo 5, Tiwanon Rd., Bangkadi, Muange 12000, Pathum Thani, Thailand
5822770525@g.siit.tu.ac.th
[3] King Mongkut's University of Technology North Bangkok,
1518 Pibulsongkram Rd., Bangsue Bangkok 10800, Thailand
s5804021620048@email.kmutnb.ac.th

Abstract. In automatic aerator-control systems used for shrimp farming, the dissolved oxygen (DO) measurement is one of the crucial parts since it affects both the quantity and quality of the product yield. It goes without saying that the more accurate the DO sensor, the more expensive it is. In this paper, we propose a technique for accuracy improvement of the DO measurement of a low-cost sensor by applying the Kalman filtering with an autoregressive model (AR). This work aims to minimize the difference between DO values read from the accurate sensor and those from less accurate sensors. Based on the standard Kalman filtering algorithm, data obtained from one low-cost sensor together with an AR of order 1 are used in the prediction stage, and data obtained from another sensor are used in the measurement update stage. Experimental results show that this technique can improve the measurement accuracy between approximately 10% and 19%.

Keywords: Dissolved oxygen measurement · Kalman filtering
Autoregressive model · Shrimp farming
Automatic aerator-control system

1 Introduction

In the past few years, efficient and intensive aqua-farming has been of interest because the world population has been increasing since the nineteenth century

© Springer Nature Switzerland AG 2019
S. Omar et al. (Eds.): CIIS 2018, AISC 888, pp. 145–156, 2019.
https://doi.org/10.1007/978-3-030-03302-6_13

and is expected to reach about 10 billion by 2050 [1]. Also, another reason is that the aqua-farming is the fastest growing food sector in the world [2]. In order to improve the performance, as well as the efficiency, of pond control systems, the embedded system technology has been introduced since three decades ago [3]. For example, a microcomputer-based control system is one of the first customized systems used to maintain large marine finfish [4]. The fundamental principle of the automatic control system for aquaculture is simple. In a nutshell, some crucial environmental parameters are monitored, and then the system uses these data to control environmental conditions according to specific requirements [5–7].

For shrimp farming, one of the most critical parameters that strongly affect the survival rate, growth rate, and gross yield is the dissolved oxygen (DO), or the oxygen content in the water [8]. Therefore, the DO measurement is one of the crucial parts of every automatic control system. It goes without saying that the more accurate the DO sensor, the more expensive it is. Recently, we have proposed an automatic aerator-control system, in which a few expensive optical DO sensors have been deployed [7]. As a consequence, the cost of the optical sensors contributes considerably to the total cost of the system. Thus, we are motivated to replace some of those expensive sensors with low-cost ones. This work aims to propose a technique that can improve the measurement accuracy of the low-cost sensor. Our proposed technique is based on Kalman filtering since it can be used to extract the signal of interest from a noisy signal.

To the best of our knowledge, the Kalman filtering has yet to apply for improving the accuracy of DO measurement in any automatic aerator-control system for shrimp farming. There are only a few related research publications. For example, Dabrowski et al. modeled water-quality parameters in prawn farming. Their proposed model does not rely on physical ecosystem modeling, and the Bayesian filtering together with the Kalman filtering is applied for inference [9]. Allen et al. investigated the use of an ensemble Kalman filter with a complex marine ecosystem model [10].

The rest of this paper is organized as follows. Section 2 provides the background information regarding our previously proposed automatic aerator-control system, the low-cost DO sensor that is developed by our team, and a summary of the Kalman filtering. Section 3 gives detail of our proposed technique. Experiment and results are shown in Sect. 4. Discussion and conclusion are made in Sects. 5 and 6, respectively.

2   Background

In this section, we first describe an overview of our previously proposed aerator-control system for shrimp farming in Thailand. Then, a concept of the low-priced, electrochemical DO sensor that is developed by our team is introduced. Last, the principle of the Kalman filtering, which is used to enhance an accuracy of the low-priced DO sensor, is briefly reviewed.

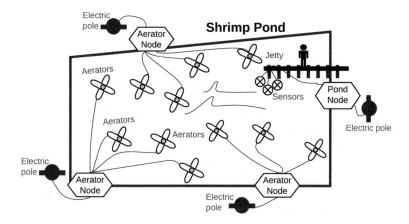

Fig. 1. Overview of the flexible and automatic aerator-control system.

2.1   Overview of an Automatic Aerator-Control System

Recently, we proposed a flexible and automatic aerator-control system for shrimp farming in Thailand [7]. It monitors some water-quality parameters, e.g., the concentration of DO and temperature, and, based on values of these parameters, aerators that have been distributed around the shrimp pond are controlled. It is a wireless sensor network consisting of one control node (the pond node) and a few remote nodes (the aerator nodes). The architecture of our system is illustrated in Fig. 1, and it operates as follows. Various sensors used for measuring water-quality parameters are attached to the pond node. Those parameter values are transmitted through the Internet and are kept in a database. Also, they are used in the calculation for making a decision about which aerators should be turned on or off including durations of operation.

The aerators are attached to the aerator nodes, which are installed around the pond, and the function of those nodes are to turn aerators on or off with respect to commands received from the pond node. The nodes communicate wirelessly via Xbee modules through the Digimesh network.

In case of emergency, e.g., a power glitch, communication failure, or a measured value is abnormal, the pond node will send an alarm message to the farmer's mobile to take an action in solving the problem. This aerator-control system can manage the pond as good as a skilled farmer and keep the pond condition suitable for shrimp living [7].

2.2   Dissolved Oxygen Sensor

One of the crucial parts in the above-described system is a degree of accuracy of DO sensors since it is what the automatic control system relies upon. It is straightforward to put that the overall effectiveness of the system can be improved if the degree of accuracy of sensors increases. However, there are at least two factors that affect the accuracy. The first one is the technology that is utilized for sensing DO. Different technologies grant different accuracy degrees. For example, the expensive optical DO sensor, which is practically used in aqua-farming, is more accurate than the relatively-cheap electrochemical one. The second factor is due to the fact that DO sensors used in automatic control systems are typically exposed to the severe environment, especially when the system monitors continuously. In this situation, the accuracy decreases with time due to the accumulation of dirt on the sensor. Inevitably, one must routinely remove dirt or biofilms from it.

In this work, we have developed the electrochemical sensor. Based on the oxygen-reduction reaction, it works as follows. When the cathode, as shown in Fig. 2, is supplied with a constant voltage, DO molecules that are diffuse through an oxygen-permeable membrane are reduced because of the oxygen-reduction action. Consequently, an electrical signal travels to the anode. We call this sensor the EST sensor after the name of our laboratory (Embedded System Technology Lab.), which is under the National Electronics and Computer Technology Center, Thailand.

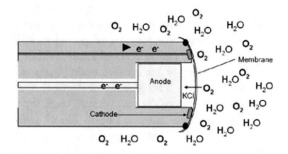

Fig. 2. Electrochemical probe utilized for sensing the concentration of DO.

Compared to an optical sensor, an electrochemical one is less accurate. However, we can improve this accuracy by applying two electrochemical sensors to acquire multichannel data. Then, the Kalman filtering is deployed to estimate a more accurate DO level. This idea is to be emphasized and explained in Sect. 2.3 and Sect. 3.

2.3   Kalman Filtering

The Kalman filter (KF) is a linear quadratic estimator that estimates the state of a linear system with random behavior. It can be used to extract or to estimate the signal of interest in a noisy signal [11–13]. The Kalman filtering algorithm consists of two stages; first, it predicts a state, and then it corrects the prediction by using measurement information. Thus, it sometimes called a predictor-corrector or a prediction-update estimator [14]. The Kalman filter model assumes that, in

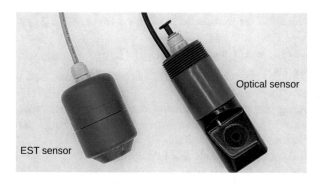

Fig. 3. EST DO sensor (which is of an electrochemical type) and optical DO sensor.

a steady state, a linear system can be described by two equations: state equation and measurement equation. The state vector $\boldsymbol{x}_k$ of the system at time k, which we intend to estimate, is evolved from the previous state at time $k-1$ according to the state equation

$$\boldsymbol{x}_k = \boldsymbol{F}_{k-1}\boldsymbol{x}_{k-1} + \boldsymbol{G}_{k-1}\boldsymbol{u}_{k-1} + \boldsymbol{w}_{k-1}, \tag{1}$$

where $\boldsymbol{F}$ is the known square state-transition matrix which applies the effect of each state parameter at time $k-1$ on the state at time k, $\boldsymbol{u}$ is the control input vector, $\boldsymbol{G}$ is the control input matrix which applies the effect of each control input parameter on the state, and $\boldsymbol{w}$ is the vector in which each element is a zero-mean process noise for each parameter in the state vector. The covariance matrix of process noise is denoted by $\boldsymbol{Q}$.

The measurement equation is described by the equation

$$\boldsymbol{y}_k = \boldsymbol{H}_k\boldsymbol{x}_k + \boldsymbol{v}_k, \tag{2}$$

where $\boldsymbol{y}_k$ is the vector of measured outputs, $\boldsymbol{H}$ is the observation matrix that relates the state vector $\boldsymbol{x}_k$ to the measurement vector $\boldsymbol{y}_k$, and $\boldsymbol{v}_k$ is the vector in which each element is a measurement white noise for each measured output. The covariance matrix of measurement noise is denoted by $\boldsymbol{R}$.

The Kalman filter algorithm consists of two stages: prediction and measurement update. The prediction stage consists of two steps as follows. Given some initial state estimate $\hat{\boldsymbol{x}}_0$ and initial state error covariance matrix $\boldsymbol{P}_0$, the state vector is first predicted from the state dynamic equation by

$$\hat{\boldsymbol{x}}_{k|k-1} = \boldsymbol{F}_{k-1}\hat{\boldsymbol{x}}_{k-1} + \boldsymbol{G}_{k-1}\boldsymbol{u}_{k-1}, \tag{3}$$

where $\hat{\boldsymbol{x}}_{k|k-1}$ is the predicted state vector (which is sometimes called *a priori* predicted state vector), and $\hat{\boldsymbol{x}}_{k-1}$ is the previous estimated state vector.

Second, the state error covariance matrix is predicted based on the following equation.

$$\boldsymbol{P}_{k|k-1} = \boldsymbol{F}_{k-1}\boldsymbol{P}_{k-1}\boldsymbol{F}_{k-1}^{\mathrm{T}} + \boldsymbol{Q}_{k-1}, \tag{4}$$

where P_{k-1} is the state error covariance of the state vector x_{k-1}, which is defined as

$$P_{k-1} = E\left[(x_{k-1} - E[x_{k-1}])(x_{k-1}^T - E[x_{k-1}^T])\right]. \tag{5}$$

Note that $E[\cdot]$ denotes the expectation of state vectors, and the superscript T denotes the matrix transposition.

After obtaining the predicted values from the prediction stage, the Kalman filtering algorithm proceeds to the measurement update stage, which consists of three steps as follows.

First, the Kalman gain matrix K_k is computed by

$$K_k = P_{k|k-1} H_k^T (H_k P_{k|k-1} H_k^T + R_k)^{-1}. \tag{6}$$

Second, the Kalman gain is used to scale a so-called "innovation," which is the difference between the measurement of the output (z_k) and the predicted output $(\hat{y}_{k|k-1} = H_k \hat{x}_{k|k-1})$, and then it is used to update the state vector, as formulated by the following equation.

$$\hat{x}_k = \hat{x}_{k|k-1} + K_k(z_k - H_k \hat{x}_{k|k-1}). \tag{7}$$

Last, the state error covariance matrix is updated by

$$P_k = P_{k|k-1} - K_k H_k P_{k|k-1}. \tag{8}$$

Note that P_k is called *a posteriori* error covariance matrix. According to the Kalman filtering algorithm, given the initial values of $\hat{x}_0$ and P_0, we can calculate the estimated state vector $\hat{x}_k$ for any time k. A diagram summarizes this algorithm is illustrated in Fig. 4.

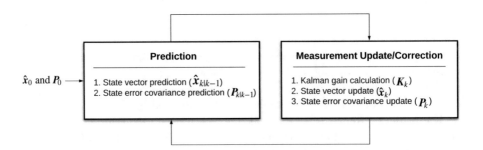

Fig. 4. Kalman filtering algorithm.

3   Proposed Method for DO Accuracy Improvement Based on Kalman Filtering

The goal of this work is to minimize the difference between DO values read from the EST and optical sensors by adopting the Kalman filtering. In our experiment, we use an autoregressive model to model the state transition of the DO concentration in the shrimp pond. Thus, data from two EST sensors are required, i.e., one is used for the prediction stage based on an autoregressive model (AR), and another is for the measurement update stage.

An autoregressive model can be used to describe a time-varying process in nature. The assumption of AR is that its output depends on its previous values and a stochastic term [15]. In general, an AR of order p is denoted by $AR(p)$ and is defined by the following equation.

$$X_t = c + \sum_{i=1}^{p} \varphi_i X_{t-i} + \varepsilon_t, \tag{9}$$

where c is a constant, φ_i for $i=1$ to p are the parameters (or coefficients) of the model, and ε_t is white noise.

In our experiment, the state transition is modeled by $AR(1)$. The framework of the Kalman filtering with $AR(1)$ is sketched in Fig. 5. It should be noted that to estimate the process covariance matrix ($\boldsymbol{Q}$) and the noise covariance matrix ($\boldsymbol{R}$), we use the smoothed data, i.e., the smoothed DO values obtained from one EST sensor, as a surrogate for the actual process stage. Also, the exponentially weighted moving average (EWMA), which is a first-order infinite impulse response (IIR) filter with exponentially-decrease weighting factors, is used to smooth the data.

In our simulation, we set the length of the moving average applying to the data to be 30 and the length of the look-back window used while performing the EWMA to be 15.

4 Experiment and Results

We conducted experiments to verify the concept of using Kalman filtering in DO measurement improvement. The details of our experiment, its setup, and results are provided in the following subsections.

4.1 Experimental Setup

In this work, we set two experiment ponds, which are cylindrical plastic containers. The diameter and the height of our experiment ponds are 1.2 m and 0.8 m, respectively. They were used to nurture 120 three-gram whiteleg shrimps (*Litopenaeus vannamei*) for 50 days, i.e., until the shrimps gained their weights of approximately 15 grams. Both ponds were in an open-air environment. The water salinity was controlled to be 17 parts per thousand, the pH was in the range of 7.5 to 8.5, and the alkalinity was in the range of 130 to 150 mg/L. The total ammoniacal nitrogen was controlled to be less than 1 mg/L.

Two EST sensors and one optical sensor, as well as a circulation pump, which was used to increase the DO content, were installed. Data obtained from those sensors were wirelessly transmitted to our database server for analysis. As

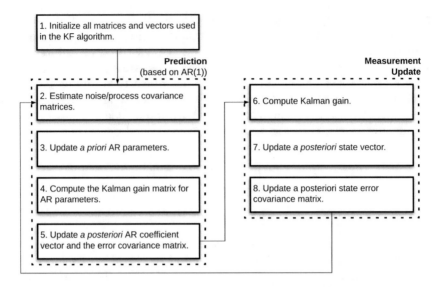

Fig. 5. Framework of the Kalman filtering with AR(1) that is used to model the state transition of the DO level.

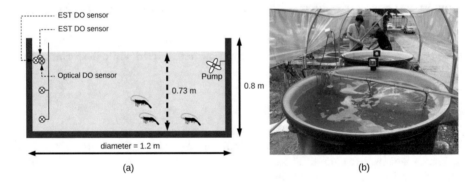

Fig. 6. Diagram of the shrimp pond together with sensors used in our experiment (left) and two experiment ponds in operation (right).

explained in the previous section, we used data from one sensor as the measurement data and those from another as the prediction model based on AR(1). Figure 6(a) shows a diagram of the experiment pond, and Fig. 6(b) shows an image of our actual ponds in operation.

4.2   Experimental Results

Figure 7 shows an example of data obtained from sensors and the proposed KF-based estimator. The comparison of the root-mean-square errors (RMSE) between DO levels obtained from the optical sensor and those obtained from the

EST sensor is shown in Table 1. Note that some data were removed from this table because some DO sensors malfunctioned during the experiment. Also, the table shows results from two situations, which are the situation in which the experiment pond has no shrimp and the situation in which the pond houses 120 shrimps.

In the former situation, i.e., there was no shrimp in the pond, when the Kalman filtering was deployed in improving the accuracy of the DO measurement, the average RMSE dropped from 0.2806 (with the standard deviation of 0.1525) to 0.2247 (with the standard deviation of 0.1142). That is, compared with using one EST sensor, the RMSE dropped approximately 19.09%. In the latter situation, the average RMSE dropped from 0.4711 (with the standard deviation of 0.3163) to 0.3959 (with the standard deviation of 0.3125), or approximately 15.96%. Therefore, in both situations, the Kalman filtering with AR(1) could improve the accuracy of the DO measurement considerably.

Table 1. Root-mean-square error (RMSE) between DO levels obtained from the optical sensor and those obtained from the EST sensor with and without the Kalman filtering.

DO-level data	Average	SD	Pond with no shrimp		Pond with 120 shrimps	
			without KF	with KF	without KF	with KF
2 − 4 July 2017	7.4081	0.6308	0.48854	0.39377	-	-
7 − 9 July 2017	7.2057	0.2168	0.12747	0.11920	-	-
10 − 12 July 2017	6.8773	0.1540	0.15781	0.12203	-	-
16 − 17 July 2017	7.0942	0.1707	0.24789	0.22012	-	-
21 − 23 October 2017	6.8083	0.0186	0.38128	0.26862	-	-
11 − 13 August 2017	6.9913	0.4625	-	-	0.93276	0.84655
14 − 16 August 2017	6.8810	0.2150	-	-	0.36677	0.22377
19 − 21 August 2017	6.5551	0.1025	-	-	0.37041	0.36048
23 − 25 August 2017	6.7844	0.0393	-		0.21438	0.15300

5   Discussion

Based on our findings, there are three issues we discuss in this section. First, even though the RMSE is reduced when the Kalman filtering is deployed, this result somehow depends upon the choice of sensor selection to be used in AR(1). That is, if we use only one sensor, instead of two, and if the data obtained from that sensor are to be used in both prediction and measurement update stages, the RMSE could either increase or decrease. However, results from our experiment confirm that, when the dynamics of DO content is modeled appropriately, this technique can be used to improve the accuracy of DO measurement.

Second, the standard deviation of the RMSE values in the situation in which the experiment pond houses 120 shrimps is about three times greater than that in another situation. This result implies that many factors affect the DO content

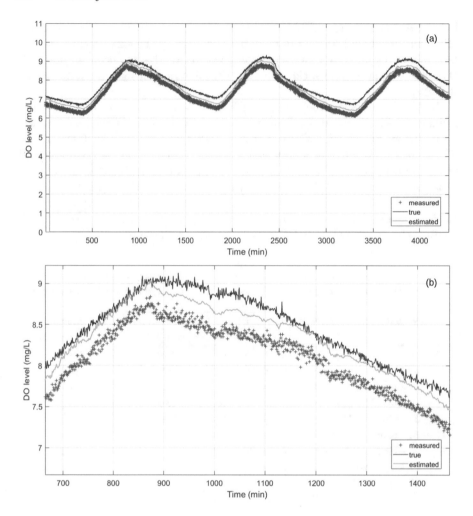

Fig. 7. Example of DO level data that are obtained from the optical sensor (blue) and EST sensor (red) and that are estimated by the proposed KF-based estimator (green): (a) data for a three-day period and (b) a close-up of the data for about 12 hours.

during nurturing and that these factors should be taken into consideration in the DO dynamic model to be used in the prediction stage of the Kalman filtering algorithm.

Last, since we did the experiment with plastic containers and neglected the effect of soil on the dissolved oxygen, the findings might not guarantee the performance or the efficiency of the technique when it is used in practical and industrial farming. Therefore, the experiment should be re-conducted in such an environment in the future.

6   Conclusion

This paper proposed a method for accuracy improvement in DO measurement by using the Kalman filtering with an autoregressive model. Instead of modeling the dynamics of the DO concentration, we used data obtained from one sensor, which were applied to AR(1), in the prediction stage of the Kalman filtering algorithm. For the measurement update stage, the data obtained from another DO sensor were used. The experimental results showed that this technique could improve the accuracy of the DO measurement considerably. Therefore, using two less accurate, but low-cost, sensors might replace a more accurate, expensive one.

Acknowledgment. This work was collaborative research between the National Electronics and Computer Technology Center (NECTEC) and the Aquaculture Product Development and Services (AAPS) laboratory of the National Center of Genetic Engineering and Biotechnology (BIOTEC), Thailand. The authors would like to express their sincere gratitude to Dr. Sage Chaiyapechara and his colleagues for domesticating the whiteleg shrimps in the experiments.

References

1. The United Nations Department of Economic and Social Affairs. http://www.un.org/en/development/desa/news/population/2015-report.html
2. The State of World Fisheries and Aquaculture. http://www.fao.org/documents/card/en/c/16c4349c-89c0-5d98-b798-922c2c2e8cae
3. Lee, P.G.: A review of automated control systems for aquaculture and design criteria for their implementation. Aquacult. Eng. **14**(30), 205–227 (1995)
4. Schlieder, R.A.: Environmentally controlled sea water systems for maintaining large marine finfish. Prog. Fish. Cult. **46**(4), 285–288 (1984)
5. Shifeng, Y., Jing, K., Jimin, Z.: Wireless monitoring system for aquiculture environment. In: Radio-Frequency Integration Technology, pp. 274–277. IEEE Press, New York (2007)
6. Cheunta, W., Chirdchoo, N., Saelim, K.: Efficiency improvement of an integrated giant freshwater-white prawn farming in Thailand using a wireless sensor network. In: Signal and Information Processing Association Annual Summit and Conference (APSIPA), pp. 1–6. IEEE Press, New York (2014)
7. Galajit, K., Duangtanoo, T., Rungprateepthaworn, K., Sartsatit, S., Dangsakul, P., Karnjana, J.: Flexible and automatic aerator-control system for shrimp farming in Thailand. In: The 2nd Advanced Research in Electrical and Electronic Engineering Technology, pp. 1–6 (2017)
8. Ghosh, L., Tiwari, G.N.: Computer modeling of dissolved oxygen performance in greenhouse fishpond: an experimental validation. Int. J. Agric. Res. **3**(2), 83–97 (2008)
9. Dabrowski, J.J., Rahman, A., George, A., Arnold, S., McCulloch, J.: State space models for forecasting water quality variables: an application in aquaculture prawn farming. In: the 24th ACM SIGKDD International Conference on Knowledge Discovery and Data Mining, pp. 177–185. ACM (2018)
10. Allen, J.I., Eknes, M., Evensen, G.: An ensemble Kalman filter with a complex marine ecosystem model: hindcasting phytoplankton in the Cretan Sea. Ann. Geophys. **21**(1), 399–411 (2003)

11. Kalman, R.E.: A new approach to linear filtering and prediction problems. J. Basic. Eng-t. ASME **82**(1), 35–45 (1960)
12. Marselli, C., Daudet, D., Amann, H.P., Pellandini, F.: Application of Kalman filtering to noisereduction on microsensor signals. In: Proceedings du Colloque interdisciplinaire en instrumentation, pp. 443–450. Ecole Normale Supérieure de Cachan (1998)
13. Lesniak, A., Danek, T., Wojdyla, M.: Application of Kalman Filter to noise reduction in multichannel data. Schedae Informaticae **17**(18), 63–73 (2009)
14. Rhudy, M.B., Salguero, R.A., Holappa, K.: A Kalman filtering tutorial for undergraduate students. Int. J. Comp. Sci. Eng. Surv. **8**, 1–18 (2017)
15. Akaike, H.: Fitting autoregressive models for prediction. Ann. I. Stat. Math. **21**(1), 243–247 (1969)

Towards Developing a Peatland Fire Prevention System for Brunei Darussalam

Nurul Wardah Haji Hamzah[(✉)], Siti Aisyah Haji Jalil[(✉)],
and Wida Susanty Haji Suhaili[(✉)]

School of Computing and Informatics,
Universiti Teknologi Brunei, Jalan Tungku Link, Gadong, Brunei
wardahhamzah@gmail.com, aisya_HJ@hotmail.com,
wida.suhaili@utb.edu.bn

Abstract. Peatland forest fire has been occurring in Belait over the Badas dome. Human activities around this area have been the main contributor. This has affected the ecosystem that leads to the degradation of the peatland forest in Brunei. These degraded areas have become prone to fire with the right conditions. To date, there are no forest fire detection devices or monitoring system installed in Brunei. This paper presents a prototype which aims to prevent peatland fire from happening. We developed this prototype using a development microcontroller board called Arduino with two types of sensors. Before we developed this prototype, we studied the requirements from the two main stakeholders of peatland forest fire: the fire fighters and the peatland experts. Taking this knowledge into account, we opted to come up with a peatland fire prevention mechanism where we artificially keep fire prone peatland areas wet during prolonged dry climate in order to reduce risk of forest fires. To facilitate this, we designed a facility consisting of two dams. In particular we use sensors that will trigger the dam valve to release water from the dam and re-wet the land when our system identifies the soil humidity level under a certain threshold value. We refer to our solution as Brunei Peatland Fire Prevention (BPFP) system. Being aware of the fact that the dams in our BPFP system could spill out during wet seasons, our system considers an automatic draining mechanism from the dams. In this paper, we demonstrate how BPFP system works. We presents the algorithms that contributes in the making of BPFP system a stable, efficient and real-time peatland fire prevention system, Additionally, in this paper, we provide a brief discussion on the major issues that would further contribute in improving our BPFP system.

Keywords: Disaster management system · Wireless sensor network
Dam · Soil moisture sensor

1 Introduction

Forest fire can cost millions of dollars in damages, affecting the environment and if the fire happens in a peatland area this will be worse. The carbon will be released and this can lead to haze which can be fatal to those with chronic breathing problems. Peat swamp forests are particularly vulnerable to fire and produce the most carcinogenic

© Springer Nature Switzerland AG 2019
S. Omar et al. (Eds.): CIIS 2018, AISC 888, pp. 157–167, 2019.
https://doi.org/10.1007/978-3-030-03302-6_14

haze of any forest type when they are burned [1]. Peat swap forests in Brunei are the second most dominant forest type in the country covering around 15.6% of the total land area [2]. The most extensive peat swamp forest is found in the Belait district and continuous with the peat swamps of the Baram over the border in Sarawak. With the new initiative of improving road in Belait district, a stretch of Peatland around Badas area was degraded. At the same time quarry of sand activities around this area has affected the ecosystem of peatland area in Belait. Hence these areas become the most occurrence of peatland fire. At the moment there is no forest fire monitoring or detection devices installed in Brunei Darussalam, thus it takes time to discover any potential forest fire. The inability to detect wild forest fire during its initial stages and to take rapid aggressive action on new fires are the most limiting factor in controlling such wild forest fires [3] and worst for peat forest as fires can start underground [4].

National Disaster Management Centre (NDMC) is Brunei Darussalam's National Focal Point for ASEAN Committee on Disaster Management (ACDM). For any occurrence of disaster in Brunei such as wild forest fires, they are the ones who will dispatch help request from the relevant assigned authorities. Forest fire incident has increase in Brunei Darussalam as year passes. Apart from prevention actions, early detection and suppression of fires is the only way to minimize the damage and losses caused by fire. Systems for early detection of forest fires have developed over the past decades based on advances in related technologies. Nevertheless despite having attempts to early detection of forest fires, the main issue still lies on the deployment issue. And the insufficient amount of water to kill the fire is another issue. Hence the next best option to manage forest fire is to stop it from happening. Since forest fire can cost millions of dollars in damages, this project explores the possibilities of having early detection system by having a prototype sensor and then transmitting data through existing web technologies focusing on the water content in the soil.

The presentation of this paper is organized in the following. The next section will compiled a related background study and discussed the pros and cons of wireless sensor network and followed by the proposed system architecture. This will then followed by the implementation of the proposed prototype, discussion and finally a conclusion is drawn.

2 Background Study

Fire resulting from peat swamp or peat fires is highly flammable, causing localized fires to spread and making them difficult to stop. Peat catches fire easily with the right conditions but peat fires are difficult to predict as they can start anywhere underground. Peat fires can go into the soil and travel underground which is dangerous for fire fighters as it can surface anywhere. Extinguishing the fire may be time-consuming since it can burn as deep as 15 feet and spread very slowly [1–4].

During dry weather, an increase in temperature can concurrently increase the chances of fires. The trees, sticks and underbrush on the ground receives radiant heat from the sun can become potential fuels for fire. This expedites the fire burning to spread swiftly.

2.1 Technologies Used for Monitoring

There have been various attempts on developing a forest fire monitoring system ranging from surveillance based, online-satellite and wireless sensor network. A variety of fire detection sensors used from video cameras, infrared thermal imaging cameras, infrared spectrometers and light detection and ranging (LIDAR) systems are also studied [5–8]. Most existing forest fire detection systems rely on the satellite imagery which includes Global Forest Watch Fire (GFW), Active Fire Mapping program and Indofire [7, 8]. However, the downside of using these systems is the high rate of false alarms due to atmospheric conditions (clouds, shadows, and dust particles), light reflection and human activities.

One inexpensive alternative to the satellite solution is by using Wireless Sensor Network (WSN) [9, 13]. This approach uses tiny, cheap and low-power sensor devices with the ability to sense the environment. They are deployed in a forest to collect the surrounding data and transmit to a processing centre through wireless broadcast medium by exchanging messages. Future risk analysis can also be done to systems that adopt WSN and using similar technology as ZigBee [11, 12].

2.2 Issues on the Monitoring and Detection System

From an interview conducted with the fire fighters and the peatland experts, it is discovered that deployment and response time are the two major issues faced if monitoring and detection system are to be created for peatland forests. Although factors including temperature, humidity of air and soil, wind speed and direction are used to detect forest fires, the deployment of such system in peatland forest is impossible. For example, the use of temperature sensors in detecting forest fires are not recommended as it only measure the ambient but the temperature of the forest environment are not monitored. Additionally, reading will be affected considering that it is taken in an open space.

The forest's safety will need to be ensured, thus a proper sensor placement needs to be considered. The use of air humidity sensor will not work as Brunei's humidity is usually high. If it is detected as low, this could mean that fire has already occurred and it is too late to react. For wind sensor, as the fire spread, wind can affect the rate of oxygen supply to the burning fuel and causing other unburned fuel to be heated [18]. Since fire can start underground, the main source of fire needs to be tackled first. Hence, soil humidity detection in the management of peatland forest fire is the best option to be used.

2.3 Soil Humidity to Monitor Forest Fire

Detection of soil humidity will ensure that the soil never dries out preventing fire to start regardless of the place and time. Several studies have verified that wet soil conditions can prevent the spread of fires. The studies emphasized that the level of surface wetness needs to be considered when assessing the risk of forest fire as well as the temporal anomalies. In [14], soil moisture data were used in wildfire danger assessment. Large growing-season wildfires occurred exclusively under low condition of soil moisture. Soil moisture was preferred as it is highly correlated to live fuel moisture to any other weather variables [15, 16].

Study on [17] has shown that the wet surface soil moisture conditions limit the extent of burned area where the area with wet soil moisture condition did not get burned. Soil moisture anomalies have been shown to be related to the occurrence of forest fires in central Siberia. Algorithms for estimating relative soil moisture have also been developed and tested from regional to a global scale [18, 19]. In the UK, the moisture content is important in combating the peat forest fires as it becomes one of the factors to be included to improve the Met Office Fire Severity Index [20].

2.4   Extinguish Fire

In peatland forest, the water collects itself naturally. Unfortunately, as mentioned earlier, the road development in the Badas area has disturbed the ecosystem. Hence this leads to dry soil and if not properly maintained will be the major cause of peatland forest fires. In Malaysia, rewetting or restoring of the level of the water table by stemming the flow of water in the canals are used to avoid drainage of peatlands and reduce fire risk. Hence the Department of Irrigation and Drainage (DID) has published a guideline on the design of check dams to raise the water levels in order to prevent peat fires [21]. Brunei fire fighters have intentions to create dams as it is rather expensive to fly water to these areas. Several firebreaks and a few dams will be created to curb the situation from reoccurring and ease the process of getting the water source. With this development, a mechanism will be implemented to provide sufficient amount of water both for dry and wet seasons.

Therefore based on these discussions, the peatland fire prevention system prototype is created. It consists of two parts: (1) it detects the humidity of the soil and (2) to ensure water source in this case from the water dam is at its sufficient level.

3   Proposed Brunei Peatland Fire Prevention (BPFP) System

The prime objective of the proposed Brunei Peatland Fire Prevention (BPFP) System is to mitigate peatland forest fire by ensuring the soil moisture level within a certain level. Our solution is devised considering the extreme peatland areas (the degraded areas where the water source has been tempered due to the heavy activities with the area, making the areas extremely fire prone). Taking account of such extreme situation, we propose to use several dams to replace the lost of water source. Assume there are N numbers of dams required in order to keep the areas wet. The value of N can be decided based on the following factors: (i) peatland area size, (ii) capacity of each dam, (iii) time required to collect water from other water sources (e.g. JKR, sea, river), (iv) the water absorption rate, (v) average soil moisture level, and (vi) availability of water supply from other natural sources (e.g. rain). It is worth highlighting that, for the prototype, we assume $N = 2$. In our proposal, one of the dams used to irrigate the soil and another used as a backup to ensure enough water supplies. The overview of our BPFP system is delineated in Fig. 1. The overall process for developing this prototype will be subsequently explained in this section.

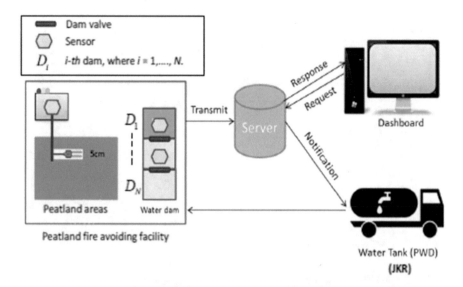

Fig. 1. System architecture of BPFP system

3.1 Overall Operational Procedure of BPFP System Prototype

Our prototype considers $N = 2$ (two dams). We consider a dashboard which retrieves sensor readings from the web-servers periodically. The web-server collects data from the sensors deployed in the peatland area (our target location). For ease of understanding the overall working procedures, we present Fig. 2, showing how the components in our prototype are arranged. Figure 2(a) portrays the block diagram of the prototype, whereas the Fig. 2(b) depicts our developed prototype at our work.

The two dams will be used to represent the two different situations where presence of water is needed. Consequently to ensure the water supply is always sufficient. A tunnel is also created to mimic the situation that will allow excess water to flow to the sea to avoid flood during rainy season (Fig. 2(a)).

During dry season, availability of water is very important. Hence *Dam1* and *Dam2* need to be filled. If the reading of the soil moisture level at time, t $\left(S_m^{(t)}\right)$ reaching below the threshold set ($S_{moisture}$), BPFP system starts pumping water of *Dam1* to the soil in order to keep the soil of the peatland fire prone area wet (i.e. $t\left(S_m^{(t)} \geq S_{moisture}\right)$ ③. And if water level (WL) in *Dam1* has a low water level under a certain threshold $\left(Th_{D1}^{low}\right)$, the *Dam2* will pump water to fill in the *Dam1* to ensure the reading of the soil is always above $S_{moisture}$ ②. If both dams have low water level, an alarm will be triggered to request for water to be filled externally to the dams. In such a situation, an SMS notification is send to Public Work Department (JKR), Brunei (this will be further explained in the subsequent part of this section). In the case where, *Dam1* has a high water level i.e. overflow, the water will be pushed back to *Dam2* (see Fig. 2(a) ④).

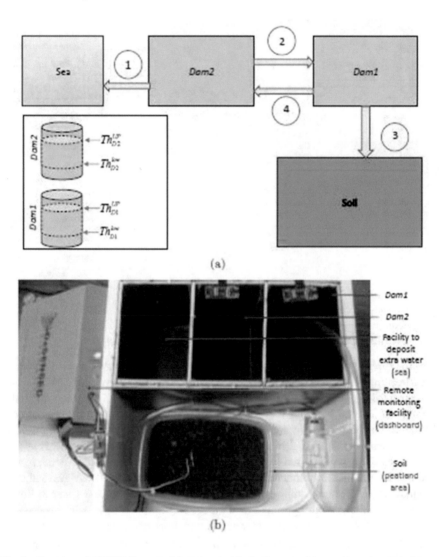

Fig. 2. Prototype for BPFP System: (a) prototype block diagram; (b) prototype developed at our work

3.2   Prototype Design and Implementation for BPFP System

The prototype is built upon a development microcontroller board called Arduino, as shown in Fig. 2(b). Arduino is an open-source platform used for building interactive electronic project. From this prototype two parts are developed: the physical system and the dashboard system.

The Arduino is connected to the internet and data is pushed to the server hosting the database, dashboard and notification. In our prototype, the physical system relies on two sensors, water pump and water gate to be in place. The two sensors used are: soil humidity sensor, which senses the moisture level and the ultrasonic sensor, which checks the water level (WL). Several actions will be triggered once a sensor sends a reading that meets a certain condition that requires a certain action. The system is also set to send alarm and able to send SMS notification. Readings will also be recorded onto a database and from the dashboard thereby allowing users to observe the current series of reading and a graph of multiple readings of the past for analysis.

The condition of each sensor plays a significant cause and effect to the overall system. Soil and dam need to be in a suitable condition to minimalize the occurrence of forest fire and ensure the cause of peatland fire is not due to an underground fire. Therefore a set of algorithms is created to represent the different conditions the system can addressed. The algorithms look at the several conditions: weather, soil, *Dam1* WL and *Dam2* WL. Based on the conditions, it makes decisions involving the water pump, tunnel, SMS alert and notification.

Before we explain our proposed algorithms which are embedded into the prototype, we defined the variables. We assume in our algorithms that each of the dams has two thresholds: lower WL and upper WL. For the two dams, this is represented as Th_{D1}^{low} and Th_{D1}^{UP} respectively for *Dam1*. While Th_{D2}^{low} and Th_{D2}^{UP} for *Dam2*.

The proposed algorithms works, first the weather is checked to see if it is wet or dry (see Fig. 3(a)). If it is dry, Algorithm 1 will run. In Fig. 3(b), Algorithm 1 checks the water level in *Dam1*. When $WL > Th_{D1}^{LOW}$ and the soil moisture is wet, this is the ideal condition. No action taken. But when the soil is dry, water valve from *Dam1* is opened to make the peatland area wet. However, if $WL \leqslant Th_{D1}^{low}$, it proceeds to check the water level in *Dam2*. If $WL > Th_{D2}^{LOW}$, water valve from *Dam2* opens to send water to *Dam1*.

On the other hand, if the weather is wet, Algorithm 2 is run. In Fig. 3(c), Algorithm 2 starts by checking the water level in *Dam1*. If $WL > Th_{D1}^{UP}$, it proceeds to check the water level in *Dam2*. If $WL \geqslant Th_{D2}^{UP}$, water from *Dam2* will flow to the sea via tunnel. But when $WL > Th_{D1}^{UP}$ and $WL > Th_{D2}^{UP}$, water valve in *Dam1* opens to send the water from *Dam1* to *Dam2*. However, when $WL \leqslant Th_{D1}^{LOW}$, water level in *Dam2* is checked. If $WL > Th_{D2}^{LOW}$, water valve in *Dam2* opens to send the water from *Dam2* to *Dam1*. But if $WL \leqslant Th_{D1}^{LOW}$ and $WL \leqslant Th_{D2}^{LOW}$, the JKR is notified via SMS to supply water to the dams.

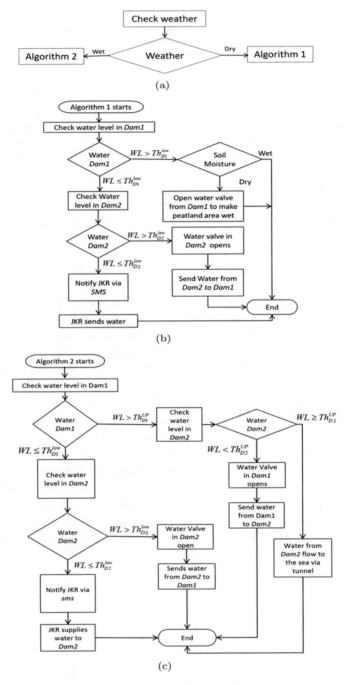

Fig. 3. Algorithms for proposed BPFP system: (a) Initial process, check weather (b) Algorithm1, weather dry; (c) Algorithm 2, weather wet.

4   Real-World Implementation Concerns

Based on the initial findings, there are four main issues that need to be addressed if this prototype is to be placed in the real environment in the future. First is the soil humidity, peat soil has a special characteristic that will need to be studied further. The moisture content in peat soil is what makes it resistant to fire. Second, the water source is mainly from the existing dams that are created. The number of dams to be used as backup reservoir and the ones used as the feeder to the soil will need to be decided. Third, for this prototype we only rely on one sensor. There will be issue of data trust which might be affected by attack whether it is deliberate example human or unintentional example animals. Hence further consideration need to be taken on how the soil moisture sensor is to be placed in the real environment. And the power source for this solution will rely on the use of solar panel. Plus to find the optimal sampling frequency data collection and transmission by the sensors can also help in reducing the power consumption. Additionally, in case when the peatland area is large, we need to make sure that all data from sensors is aggregated properly and delivered to the web-server with desired Quality of Service (QoS).

5   Conclusion

The prototype addresses the requirements stated by the two main stakeholders of peatland forest fire. The water control mechanism installed for the small dam would ensure the soil is sufficiently wet and reading should not go beyond the set assumed threshold. At the same time the dedicated use of dams ease the task of getting water to the affected areas during dry season. Issues of deployment will no longer exist as the development of the prototype goes in line with the requirements set by the two stakeholders. This proof of concept can be applied in Brunei Darussalam. With the advancement of technology, an innovative preventive method of monitoring the soil humidity and controlling the water level of the dam is made possible to minimize the occurrence of forest fire in Brunei. Furthermore this prototype system works on web dashboard system via wireless technologies so that it helps the Brunei Fire and Rescue department to monitor real time environmental contexts of the soil mixture and water level of the dams. This solution to peatland fire management will contribute to the effective means of protection and conserving peatland forest from further degradation due to human activities such as road improvement or agriculture. It has provided a good opportunity for undergraduate research experience and can be further extended to a larger scale. In the next project the expansion of soil humidity reading need to be verified such as taking readings from various depths to ensure a true reading [22] as well as attending the highlighted issues in the implementation concern section.

References

1. Pearce, G.: The Science of fire behavior and fire danger rating, Tech Rep, New Zealand Forest Research Institute Ltd. (2000)
2. Forest Department Brunei. http://forestry.gov.bn/SitePages/PeatSwamp.aspx
3. Roy, P.S.: Forest fire and degradation assessment using satellite remote sensing and geographic information system. Satellite Remote Sensing and GIS Application in Agricultural Meteorology, 361–400 (2014)
4. Tishkov, A.A.: Sub-Surface Peat Fires, Natural Disasters. Encyclopedia of Life Support Systems (EOLSS), vol. 11 (2010)
5. Rego, F.C., Catry, F.X.: Modelling the effects of distance on the probability of fire detection from lookouts. Int. J. Wildland Fire **15**(2), 197–202 (2006)
6. Stipanicev, D., Vuko, T., Krstinic, D., Stula, M., Bodrozic, L.: Forest fire protection by advanced video detection system. In: Proceedings of the Third TIEMS Workshop-Improvement of Disaster Management System, pp. 26–27 (2006)
7. Utkin, A.B., Fernandes, A.M., Simoe, F., Lavory, A., Vilar, R.: Feasibility of forest-fire smoke detection using LIDAR. Int. J. Wildland Fire **2**(2), 159–166 (2006)
8. Lavrov, A., Utkin, A.B., Vilar, R., Fernandes, A.M.: Application of LIDAR in ultraviolet, visible and infrared ranges for early forest fire detection. Appl. Phys. B-Lasers Optic **76**(1), 87–95 (2003)
9. Lozano, C., Rodriguez, O.: Design of forest fire early detection system using wireless sensor networks. Online J. Electron. Electr. Eng. **3**(2), 402–405 (2011)
10. Hafeeda, M., Bagheri, M.: Forest fire modeling and early detection using Wireless Sensor Networks. Adhoc Sens. Wirel. Netw. **7**, 169–224 (2009)
11. Nadhi, P., Hiren, K., Arjaz, B.: Wireless sensor network using Zigbee. Int. J. Res. Eng. Technol. **2**(6), 10381042 (2013)
12. Jadhav, P.S., Deshmukh, V.U.: Forest fire monitoring system based on ZIGBEE wire-less sensor network. Int. J. Emerg. Technol. Adv. Eng. **2**(12), 187–191 (2012)
13. Majorne, B., Viani, F., Filippoi, E., Bellin, A., Massa, A., Toller, G., Robol, F., Saluci, M.: Wireless Sensor Network deployment for monitoring soil moisture dynamics at the field scale. Procedia Environ. Sci. **19**, 426–435 (2013)
14. Krueger, E.S., Ochsner, T.E., Engle, D.M., Carlson, J.D., Twidwell, D.L., Fuhlendorf, S.D.: Soil Moisture affects growing-season Wildfire Size in the Southern Great Plains, Agronomy & Horticulture, Faculty Publication (2015). https://digitalcommons.unl.edu/agronomyfacpub/838
15. Pelizzaro, G., Cesaraccio, C., Duce, P., Ventura, A., Zara, P.: Relationships between seasonal patterns of live fuel moisture and meteorological drought indices for Mediterranean shrubland species. Int. J. Wildland Fire **16**, 232–241 (2007). https://doi.org/10.1071/WF06081
16. Qi, Y., Dennison, P., Spenser, J., Riano, D.: Monitoring live fuel moisture using soil moisture and remote sensing proxies. Fire Ecol. **8**, 71–87 (2012). https://doi.org/10.4996/fireecology.0803071
17. Bartsch, A., Balzter, H., George, C.: Influence of regional surface soil moisture anomalies on forest fires in Siberia observed from satellites. Environ. Res. Lett. **4**, 405021 (2009)
18. Wolfgang, W., Lemoine, G., Helmut, R.: A method for estimating soil moisture from ERS scatterometer and soil data. Remote Sens. Environ. **70**, 191–207 (1999)
19. Wolfgang, W., Vahid, N., Klaus, S., de Jeu, R., Martinez-Fernandez, J.: Soil moisture from operational meteorological satellites. Hydrol. J. **15**, 121–131 (2007)

20. Krivtsov, V., Gray, A., Valor, T., Legg, C.J., Davies, G.M.: The fuel moisture content of peat in relation to meteorological factors. Modelling, Monitoring and Management of Forest Fire I. WIT Trans. Ecol. Environ. **119**, 193–200 (2008)
21. Forestry Department Peninsular Malaysia, GAMBUT, ASEAN Peatland Forests Project Malaysia Component, Special Publication (2014)
22. Ritzema, H., Limin, S., Kusin, K., Jauhiainen, J., Wosten, H.: Canal blocking strategies for hydrological restoration of degraded tropical peatlands in Central Kalimantan, Indonesia. Catena **111**, 11–20 (2013)

Smartwatch-Based Application for Enhanced Healthy Lifestyle in Indoor Environments

Gonçalo Marques and Rui Pitarma$^{(\boxtimes)}$

Polytechnic Institute of Guarda – Unit for Inland Development, Av. Dr. Francisco Sá Carneiro, nº 50, 6300-559 Guarda, Portugal
goncalosantosmarques@gmail.com, rpitarma@ipg.pt

Abstract. A productive and healthy environment is directly influenced by indoor air quality parameters. Therefore, is fundamental to monitor indoor air quality environments as in a great diversity of living environments the air quality can be extremely poor. Humans typically spend more than 90% of the time indoors thus is extremely important to detect air quality problems in real-time. The unceasing scientific developments turn achievable to develop systems alongside with data collection and data sharing leading to several enhancements in ambient assisted living systems architectures. In this paper, a smartwatch-based application for enhanced living environments based on Internet of Things is presented. This system incorporate a hardware prototype for data sensing denominated by iAQ Wi-Fi and a smartwatch application that provides data consulting and notifications. The iAQ Wi-Fi incorporate wireless communication technologies and offers modularity, scalability, easy installation and smartwatch compatibility. The real-time monitoring data is stored in a cloud service named ThingSpeak and can be accessed by a smartwatch application denominated by iAQ Watchapp which allow easier access to the living environment quality in real time. Using the iAQ Watchapp the regular can analyze the monitored data in numeric or chart form.

Keywords: Ambient Assisted Living · e-Health
Enhanced living environments · Internet of Things · Smartwatch

1 Introduction

Ambient Assisted Living (AAL) is a research field that aims to offer an environment of different kinds of sensors, processors, mobile, wireless networks and software solutions for enhanced living environments [1]. Nowadays, there are different AAL architectures based on numerous sensors for measuring physiological parameters, environmental conditions and geolocation which regularly use wireless communication technologies such as Bluetooth-based technologies, Wifi-based technologies, NFC (Near Field Communication)-based technologies and GSM-based technologies.

There are several open issues not only in designing and development of efficient AAL solutions such as data architecture, interface design, human-computer communication, ergonomics, usability and availability [2] but also social and moral difficulties such as the adoption by the older adults. Is also crucial to guarantee that technology

© Springer Nature Switzerland AG 2019
S. Omar et al. (Eds.): CIIS 2018, AISC 888, pp. 168–177, 2019.
https://doi.org/10.1007/978-3-030-03302-6_15

does not replace human care while it must be used as a significant compliment. At the second half of this century, 20% of the humankind will be age 60 or above [3] that will result in a rise of syndromes, health care costs, scarcity of caregivers that lead to a significant social effect.

Indoor air quality (IAQ) is an important factor of personal exposure to contaminants as regular people expend about 90% of their time in indoor environments. Concerning the older people and new-borns who are very liable disturbed by this contaminants may spend all their time in indoors [4].

Indoor and outdoor air quality is normalised by Environmental Protection Agency (EPA). This agency had positioned air quality in top 5 environmental risks to the public health as indoor levels of contaminants can be up to 100 times greater than outdoor pollutant level [5].

Therefore, the iAQ Watchapp a smartwatch-based application for IAQ supervising based on the Internet of Things (IoT) was created by the authors for enhanced occupational health. This architecture incorporate not only a hardware prototype responsible for data sensing and acquisition but also smartwatch-based app data access and notifications.

2 Wearable Sensors and Indoor Air Quality

Wearable sensors are used in a large scale for AAL. The wearable sensors can be used linked to data treatment algorithms and visualisation tools to monitor real-life environments and physiological parameters.

The smartwatch incorporates a wide range of sensors and wireless communication technologies that can be used to create ubiquitous and pervasive health monitoring solutions. The Fig. 1 represents the wearable sensors incorporated in a smartwatch.

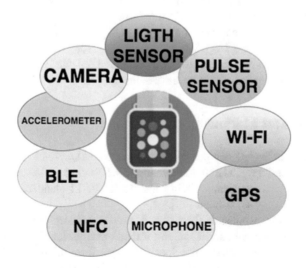

Fig. 1. Smartwatch incorporated sensors

The review of state of the art shows us intelligent systems and projects that lead us to the assertion of wearable sensors as having a significant role in the future and presents of the AAL systems.

Wearable motion sensors can help to detect behavioural anomaly situation in smart AAL by collect motion data combined with locational context [6].

Wearable technologies can be used to decrease the healthcare cost by allowing people to be supervised in their own houses, rather than in hospitals, for a portion of the price [7]. Wristbands are used as non-invasive sensors for measuring and monitoring various physiological parameters such as ECG (electrocardiograph), EEG (electroencephalograph), EDA (electrodermal activity), respiration, and even biochemical processes such as wound healing [8]. A smartwatch interface for communication and notification for hospitals that incorporates call light system, chair and bed alerts, wander guard, and emergency call purposes to provide helpful alarms for nursing staff is proposed by [9].

Scientific projects must remain to study IAQ complications and adapt legislation and inspection mechanisms to take actions in real time to increase occupational health both in public or private places by creating more rigid rules for the construction of buildings. In the majority of the cases, easy interventions that can be done by homeowners and building managers can generate helpful effects in IAQ. The inhibition to smoke inside buildings and the usage of natural ventilation are indispensable behaviours that must be communicated to youngsters using educational programs for enhanced IAQ [10].

IAQ is essential for public health. Although, there is a scarcity of attention in the scientific study of ways to increase IAQ in industrialised nations [11]. IAQ real-time supervision is as an indispensable instrument for enhanced living environments as it allows to plan interventions to improve occupational health.

Several IoT solutions for air quality monitoring that use open-source technologies for real-time data processing, collection and sharing are proposed by [12–22].

Associating the importance of IAQ monitoring with the advantages of wearable technologies the authors developed the iAQ Watchapp. This smartwatch app provides fast and easy access to the IAQ data to the regular user. The iAQ Watchapp allow ubiquitous and pervasive access to IAQ data and enable real-time detection of unhealthy conditions in indoor living environments.

3   Materials and Methods

The IAQ data is collected using a prototype named iAQ Wi-Fi which incorporates luminosity, temperature, CO_2, humidity and PM10 sensors. It uses the ESP8266 module for Wi-Fi communication using the IEEE 802.11 standard that offers radio diffusion within the 2.4 GHz band [23].

The IAQ Wi-Fi is responsible for data collection and incorporates an Arduino UNO [24] as a processing unit and an ESP8266 for data transmission. The monitored data is stored in a cloud service named ThingSpeak. This platform offers an API (Application Programming Interface) to store and retrieve data using HTTP [25].

In order to access the monitored data, the regular user can access the ThingSpeak Web portal or use the smartwatch application created in JAVA using the IDE (Integrated Developing Environment) Android Studio.

The iAQ Watchapp is implemented using the API 23: Android 6.0. The tests are conducted in a Sony Smartwatch 3 SWR50, that features a Quad ARM A7, 1,2 GHz, 512 MB de RAM and an eMMC de 4 GB, it also supports Wi-Fi, NFC (Near Field Communication) and Bluetooth. This smartwatch application provides not only a history of changes to help the detailed analysis of IAQ parameters but also support decisions in possible interventions for enhanced living environments.

The iAQ Wi-Fi solution is based on IoT [26] a paradigm that introduces the concept of the ubiquitous presence of a diversity of entities which can be identified by the unique address. The system architecture of that solution is represented in Fig. 2. To allow ubiquitous and pervasive access to IAQ data and enable real-time detection of unhealthy conditions in indoor environments iAQ Watchapp a smartwatch-based application was created by the authors.

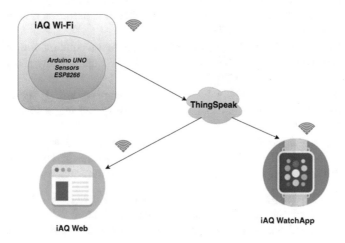

Fig. 2. iAQ Wi-Fi system architecture.

The iAQ Wi-Fi incorporate an Arduino UNO microcontroller. The Arduino is an open-source computer and software platform that uses an Atmel AVR CPU [24] as a processing unit. For data collection, the iAQ Wi-Fi incorporate several sensors such as a MG-811 CO_2 sensor, humidity and temperature sensor, a dust sensor and TSL2561 luminosity sensor. The ESP8266 is responsible for the wireless communication unit.

The iAQ Wi-Fi hardware that is responsible for environmental data collection is represented in Fig. 3, a report of the sensors used is presented.

- **TH2** – is a I2C temperature and relative humidity sensor. This sensor can measure relative humidity with a range of 0–80% and an accuracy of ±4.5%. The temperature range is 0–70 °C with an accuracy of ±0.5 °C.

Fig. 3. iAQ Wi-Fi prototype

- **Shinyei Model PPD42NS** – is a PWM sensor that uses a counting method to measure dust concentration. It has reliable and sensitive detection as it is responsive to a PM of diameter 1 μm [27].
- **TSL2561** – is a I2C digital light sensor that has a high resolution of a 16-Bit digital output and range of 0.1–40,000 LUX. The TSL2561 can operate at a temperature range of −40 °C to 85 °C [28].
- **MG-811** – is a CO2 sensor which is extremely responsive to CO_2 and less sensitive to CO and alcohol. The MG-811 have low humidity and temperature dependence [29, 30].
- **ESP8266** – is a Wi-Fi chip with a built-in antenna and support 802.11 b/g/n protocols, Wi-Fi 2.4 GHz, WPA/WPA2, incorporate 32 bits CPU with a 80/160 MHz clock speed and can operate at temperature range −40C–125C [31].

iAQ Wi-Fi prototype firmware was been developed using the Arduino IDE. The Arduino UNO and the ESP8266 communicate by a serial RS232. The Arduino microcontroller send the data collected by the sensors to ESP8266. Furthermore, the ESP8266 to upload this data to the cloud using Thingspeak platform these data can be accessed by the smartwatch application. The configuration of the Wi-Fi network to which iAQ Wi-Fi will be connected can be done by the regular user and supports auto-configuration. The ESP8266 can be set to hotspot mode to configure the network SSID and password. The smartwatch communication can be done not only by using BLE (Bluetooth 4.0 Low Energy) using a smartphone but also provides Wi-Fi direct Internet connectivity (Fig. 4).

4   Results and Discussion

The iAQ Watchapp offers a quick and reliable access to the monitored data using a numerical values or chart form. A sample of the functionalities of this smartphone-based application is shown in Figs. 6, 7 and 8. The smartwatch application homepage is represented by Fig. 5.

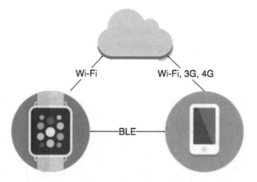

Fig. 4. Smartwatch data access

Fig. 5. iAQ Watchapp homepage

The user can see the type of data, the data and the collection timestamp but also an emotion that gives him simple and intuitive feedback to parameter quality.

The applications have two buttons, one to change the parameter type (temperature, humidity, luminosity, CO2 and dust) and the other button to show the values in a chart form (Fig. 6).

Fig. 6. Application workflow

Figure 7 shows the graphics display functionality of the smartwatch application. The graphs values represent relative humidity values measured in % (A), luminosity data measured in lux (B), dust concentration measured in µg/m3 (C) and temperature data measured in Celsius (D).

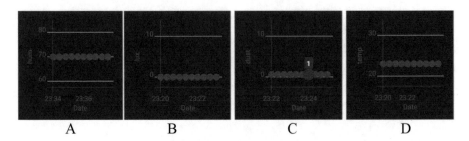

A   B   C   D

Fig. 7. Graphics display functionality

Smartwatch enables to receive unobtrusive notifications in a great diversity of situations and also could reduce smartphone dependency [32]. In United States the smartphone adoption has increased from 33% in 2011, 56% at the end of 2013 and 64% in early 2015 [33]. Smartwatches incorporate significant processing and storage features that make possible to the regular user an easy, ubiquitous, fast, and intuitive access to IAQ parameters of their home.

As referred by [34] emoticons in instant messenger software's stand as samples of expressively semantic communication in the area of on-screen interface. The iAQ Watchapp uses emoticons to provide in a ubiquitous way the current information about the indoor health living environment. The Fig. 8 shows the use of the emoticons in the smartwatch application, this functionality wishes to offer understanding information to the end user who does not require technical knowledge about IAQ.

Fig. 8. iAQ Wi-Fi mobile app

The graphics view offers a sophisticated observation of the IAQ behaviour than the numerical view format. The smartwatch application additionally offers an easy and quick to access the monitored parameters, which allows a detailed analysis of the IAQ evolution.

Consequently, the iAQ Wi-Fi can be assumed as an important method for IAQ evaluation and to plan interventions to increase a healthy and more productive indoor living environment. Assumed the ubiquity of smartwatches, this enhanced user interface is an ideal solution for wireless health applications with low user burden [35].

In spite of all the advantages in the use of smartwatch two import issues still existing such as the small size of the display and the battery capacity. Therefore, the screen interface must provide an efficient cognitive message and must be directly understood by the user [36]. Another important issue for smartwatch usage is as referred by [37] it is not yet clear whether devices like smartwatches will become all pervasive in regions—such as the UK and USA—where smartphones are dominant.

The smartwatches are a ubiquitous and pervasive method to visualise notifications; the wrist is an amazing location to notice changes in the air quality to the user in an efficient and fast modus [38].

The iAQ Watchapp is at least one of the firsts smartwatch applications that aims to improve the air quality for enhanced living environments. As described and concluded by several studies the IAQ is directly related to health quality and productivity. Compared to other systems the iAQ Watchapp provides a ubiquitous and pervasive solution for IAQ that permits the regular user to take interventions to increase the quality of their indoor environment in real-time.

5 Conclusion

IAQ monitoring could improve occupant health lifestyle and analyse the state of ventilation and the real health conditions of living environments in real time. The smartwatch applications provide ubiquitous access to living environments data; real-time notifications permit the immediate intervention of the end user to increase the IAQ in indoor environments. Therefore, this paper had described a ubiquitous and pervasive smartwatch-based solution for IAQ monitoring. This system incorporates a prototype for indoor data acquisition and smartwatch-based application for data accessibility and notifications.

The results achieved are promising, indicating an essential contribution to IAQ supervision platforms. The solution has numerous benefits compared to other solutions such as the incorporation of cost-effective, open-source technologies, compatibility with all buildings and smartwatch compatibility.

System improvements are planned to adjust the solution to particular cases such as colleges and hospitals but likewise several features enhancements as well as the creation of communication methods to offer data sharing in a safe approach to health specialists.

The authors consider that indoor environments will implement air quality monitoring because it is particularly valuable to deliver support to a medical evaluation by clinical specialists in a near future.

References

1. Universal Open Platform and Reference Specification for Ambient Assisted Living. http://www.universaal.org/
2. Koleva, P., Tonchev, K., Balabanov, G., Manolova, A., Poulkov, V.: Challenges in designing and implementation of an effective Ambient Assisted Living system. In: 2015 12th International Conference on Telecommunication in Modern Satellite, Cable and Broadcasting Services (TELSIKS), pp. 305–308 (2015)
3. UN, 'Worldpopulationageing: 1950–2050,' pp. 11–13 (2001)
4. Walsh, P.J., Dudney, C.S., Copenhaver, E.D.: Indoor Air Quality. CRC Press, USA (1983)
5. Seguel, J.M., Merrill, R., Seguel, D., Campagna, A.C.: Indoor air quality. Am. J. Lifestyle Med. **11**(4), 284–295 (2016)
6. Zhu, C., Sheng, W., Liu, M.: Wearable sensor-based behavioral anomaly detection in smart assisted living systems. IEEE Trans. Autom. Sci. Eng. **12**(4), 1225–1234 (2015)
7. Aced López, S., Corno, F., De Russis, L.: Supporting caregivers in assisted living facilities for persons with disabilities: a user study. Univers. Access Inf. Soc. **14**(1), 133–144 (2015)
8. Acampora, G., Cook, D.J., Rashidi, P., Vasilakos, A.V.: A survey on ambient intelligence in health care. Proc. IEEE Inst. Electr. Electron. Eng. **101**(12), 2470–2494 (2013)
9. Ali, H., Li, H.: Designing a smart watch interface for a notification and communication system for nursing homes. In: Zhou, J., Salvendy, G. (eds.) Human Aspects of IT for the Aged Population. Design for Aging, vol. 9754, pp. 401–411. Springer, Cham (2016)
10. Jones, A.P.: Indoor air quality and health. Atmos. Environ. **33**(28), 4535–4564 (1999)
11. Sundell, J.: On the history of indoor air quality and health. Indoor Air **14**(s7), 51–58 (2004)
12. Pitarma, R., Marques, G., Ferreira, B.R.: Monitoring indoor air quality for enhanced occupational health. J. Med. Syst. **41**(2), 23 (2017)
13. Marques, G., Pitarma, R.: Health informatics for indoor air quality monitoring. In: 2016 11th Iberian Conference on Information Systems and Technologies (CISTI), pp. 1–6 (2016)
14. Pitarma, R., Marques, G., Caetano, F.: Monitoring indoor air quality to improve occupational health. In: Rocha, Á., Correia, A.M., Adeli, H., Reis, L.P., Mendonça Teixeira, M. (eds.) New Advances in Information Systems and Technologies, vol. 445, pp. 13–21. Springer, Cham (2016)
15. Marques, G., Pitarma, R.: Smartphone application for enhanced indoor health environments. J. Inf. Syst. Eng. Manag. **4**(1), 9 (2016)
16. Marques, G., Pitarma, R.: Monitoring health factors in indoor living environments using Internet of Things. In: Rocha, Á., Correia, A.M., Adeli, H., Reis, L.P., Costanzo, S. (eds.) Recent Advances in Information Systems and Technologies, vol. 570, pp. 785–794. Springer, Cham (2017)
17. Marques, G., Pitarma, R.: Monitoring and control of the indoor environment. In: 2017 12th Iberian Conference on Information Systems and Technologies (CISTI), pp. 1–6 (2017)
18. Feria, F., Salcedo Parra, O.J., Reyes Daza, B.S.: Design of an architecture for medical applications in IoT. In: Luo, Y. (ed.) Cooperative Design, Visualization, and Engineering, vol. 9929, pp. 263–270. Springer, Cham (2016)
19. Ray, P.P.: Internet of Things for smart agriculture: technologies, practices and future direction. J. Ambient Intell. Smart Environ. **9**(4), 395–420 (2017)
20. Matz, J.R., Wylie, S., Kriesky, J.: Participatory air monitoring in the midst of uncertainty: residents' experiences with the speck sensor. Engag. Sci. Technol. Soc. **3**, 464 (2017)
21. Demuth, D., Nuest, D., Bröring, A., Pebesma, E.: The AirQuality SenseBox. In: EGU General Assembly Conference Abstracts, vol. 15 (2013)

22. Marques, G., Roque Ferreira, C., Pitarma, R.: A system based on the Internet of Things for real-time particle monitoring in buildings. Int. J. Environ. Res. Public Health **15**(4), 821 (2018)
23. Bhoyar, R., Ghonge, M., Gupta, S.: Comparative Study on IEEE Standard of Wireless LAN/Wi-Fi 802.11 a/b/g/n. Int. J. Adv. Res. Electron. Commun. Eng. IJARECE **2**(7), 687–691 (2013)
24. D'Ausilio, A.: Arduino: a low-cost multipurpose lab equipment. Behav. Res. Methods **44**(2), 305–313 (2012)
25. Doukas, C., Maglogiannis, I.: Bringing IoT and Cloud Computing towards Pervasive Healthcare, pp. 922–926 (2012)
26. Atzori, L., Iera, A., Morabito, G.: The Internet of Things: a survey. Comput. Netw. **54**(15), 2787–2805 (2010)
27. Austin, E., Novosselov, I., Seto, E., Yost, M.G.: Laboratory evaluation of the Shinyei PPD42NS low-cost particulate matter sensor. PLoS ONE **10**(9), e0137789 (2015)
28. Minghui, Y., Peng, Y., Wangwang, S.: Light intensity sensor node based on TSL2561. Microcontrollers Embed. Syst. **6**, 017 (2010)
29. da Lima, A.L., da Silva, V.L.: Micro sensor para monitoramento da qualidade do ar. In: Workshop de Gestão, Tecnologia Industrial e Modelagem Computacional, vol. 1 (2015)
30. Banick, J.L., Zolkowski, J.J., Lenz, K.E., Sanders, J.: Monitoring carbon dioxide and methane levels above retired landfill and forest control site with a tethered aerostat to determine remediation effectiveness. In: Proceedings of the Wisconsin Space Conference (2016)
31. Espressif Systems: ESP8266EX Datasheet (2015). http://download.arduino.org/products/UNOWIFI/0A-ESP8266-Datasheet-EN-v4.3.pdf
32. Cecchinato, M.E., Cox, A.L., Bird, J.: Smartwatches: the good, the bad and the ugly?, pp. 2133–2138 (2015)
33. Müller, H., Gove, J.L., Webb, J.S., Cheang, A.: Understanding and comparing smartphone and tablet use: insights from a large-scale diary study. In: Proceedings of the Annual Meeting of the Australian Special Interest Group for Computer Human Interaction, pp. 427–436 (2015)
34. Ross, P.R., Overbeeke, C.J., Wensveen, S.A.G., Hummels, C.M.: A designerly critique on enchantment. Pers. Ubiquitous Comput. **12**(5), 359–371 (2008)
35. Kalantarian, H., Alshurafa, N., Nemati, E., Le, T., Sarrafzadeh, M.: A smartwatch-based medication adherence system, pp. 1–6 (2015)
36. Lutze, R., Waldhor, K.: A smartwatch software architecture for health hazard handling for elderly people, pp. 356–361 (2015)
37. Jones, M., et al.: Beyond 'yesterday's tomorrow': future-focused mobile interaction design by and for emergent users. Pers. Ubiquitous Comput. **21**(1), 157–171 (2017)
38. Kerber, F., Hirtz, C., Gehring, S., Löchtefeld, M., Krüger, A.: Managing smartwatch notifications through filtering and ambient illumination, pp. 918–923 (2016)

A Data Mining Approach for Inventory Forecasting: A Case Study of a Medical Store

Burhan Rashid Hussein$^{(\boxtimes)}$, Asem Kasem, Saiful Omar, and Nor Zainah Siau

School of Computing and Informatics,
Universiti Teknologi Brunei, Gadong, Brunei
burhr2@gmail.com,
{asem.kasem, saiful.omar, zainah.siau}@utb.edu.bn

Abstract. One of the factors that often result in an unforeseen shortage or expiry of medication is the absence of, or continued use of ineffective, inventory forecasting mechanisms. Unforeseen shortage of perhaps lifesaving medication potentially translates to a loss of lives, while overstocking can affect both medical budgeting as well as healthcare provision. Evidence from literature indicates that forecasting techniques can be a robust approach to address this inventory management challenge. The purpose of this study is to propose an inventory forecasting solution based on time series data mining techniques applied to transactional data of medical consumptions. Four different machine learning algorithms for time series analysis were explored and their forecasting accuracy estimates were compared. Results reveal that Gaussian Processes (GP) produced better results compared to other explored techniques (Support Vector Machine Regression (SMOreg), Multilayer Perceptron (MLP) and Linear Regression (LR)) for four weeks ahead prediction. The proposed solution is based on secondary data and can be replicated or altered to suit different constraints of other medical stores. Therefore, this work evidently suggests that the use of data mining techniques could prove a feasible solution to a prevalent challenge in medical inventory forecasting process. It also outlines the steps to be taken in this process and proposes a method to estimate forecasting risk that helps in deploying obtained results in the respective domain area.

Keywords: Medical inventory forecasting · Time-series forecasting
Medical stores · Data mining

1 Introduction

Demand forecasting has become inevitable as it is one of the key areas in supply and chain management. Different studies have been conducted on demand forecasting by time series data mining techniques such as forecasting demand for electricity, tourism, food product sales, and other types of products and services [1, 4, 5]. In pharmaceutical industry, as one of the vital healthcare sectors, it is critical to ensure sufficient availability of medicines when demanded as well as keeping the spending within allocated budget [4]. Medical shortage and overstocking have been a challenge in most of the pharmaceutical companies. Pharmaceutical demand forecasting is a complex task that

© Springer Nature Switzerland AG 2019
S. Omar et al. (Eds.): CIIS 2018, AISC 888, pp. 178–188, 2019.
https://doi.org/10.1007/978-3-030-03302-6_16

involves several factors, such as seasonal variations, pharmacologic category and application of medicines, geographical diversity of consumers with respect to the therapy used by practitioners, being a new product and drug prices [2]. It has been reported that the use of statistical formulas that are usually supplied by WHO to this industry are becoming more complex and hence impractical to implement in certain countries. By utilizing demand records of previous years, data mining techniques try to provide an alternative solution for this challenge.

In this study, we proposed a data mining solution that can be used by a medical store to forecast the demand of medical and pharmaceutical products or items. The solution utilizes time series models together with error estimation using Root Mean Square Error (RMSE) to enable responsible officers/practitioners in the medical store to employ the forecasting results. Four machine learning algorithms for time series analysis have been considered in this study, namely Linear Regression (LR), Multilayer Perceptron (MLP), Gaussian processes (GP) and Support Vector Machine Regression (SMOreg) and the evaluation results have been discussed. The proposed solution, however, can be consumed using other time series models, and it can be tailored to suit medical demand forecasting based on different store constraints.

The next section provides a concise overview of the literature review on various applications of data mining techniques in demand/inventory forecasting. Section 3 presents the proposed approach and how to assess for risk in placing orders based on forecasted results. Section 4 discusses our experimental results, and we finally provide conclusions about our study in Sect. 5.

2   Related Works

Time series data mining techniques allow machines to examine large records of time varying data to build forecasting models, which is a usual case with medical store inventory records. Machine learning algorithms have been widely used in building prediction models and in time series forecasting [3]. In some prior studies, these algorithms have been widely adopted as classical methods for forecasting data with nonlinear behaviors [5].

In one study conducted by Zadeh [6], data mining was applied to determine the quantity of drug to be kept in store. First, explorative network analysis was done to discover clique set and group members of drugs with similar sales behavior. The study used three different time series sales forecasting models. A linear model (ARIMA) and a hybrid neural network model was firstly applied to individual drugs past records to build a sales forecasting model. Finally, a hybrid neural network model was built by using each drug's own past records together with co-member drugs of similar cluster from exploratory analysis. The latter approach proved superior compared to the first two approach as it helped in capturing both linear and non-linear trends of drug sales.

Another study [7] was conducted to determine the accuracy of other forecasting techniques in the inventory planning of a pharmaceutical company, which were Moving Average, Exponential Smoothing, and Winter's Exponential Smoothing. The company reported a shortage and out of stock cases in certain business days. According to the preliminary survey conducted by the study, it was revealed that the importance of forecasting is underestimated in their supply chain process. Two products were empirically chosen; one with a stable trend and another with a seasonal trend. It was concluded that six-month Moving Average predicted the product with stable demand more accurately, while Winter's Exponential Smoothing produced more accurate results in predicting the seasonal product.

In another study [8], a hybrid forecasting phase which combines Moving Average model and Bayesian Network approach was used. The study clusters product distribution warehouses similar sales behavior using bipartite graph clustering. The process involves four stages; data preparation for applying data mining algorithm; Moving Average value was calculated for each product; Clustering was applied to group warehouse and sub-distribution with similar sales behavior; and finally, Bayesian network was applied to obtain forecasting results. Experimental results on a real dataset showed that the suggested approach improves forecasting performance compared to the primitive forecasting method (Moving average).

A different approach was used in another study [9], which focused more on the out of stock prediction at product/store level rather than the quantity of stock. The first phase of the modelling process consists of clustering stores in the supply chain based on aggregate sales patterns. After store-cluster models were constructed, these clusters were used to more accurately make out-of-stock predictions at the store/product level using decision trees and neural network algorithms. The paper suggested using time series forecasting if there is sufficient temporal data.

Apart from inventory applications, time series data mining techniques such as Artificial Neural Network (ANNs) has outperformed different classical statistical methods such as linear regression and Box-Jenkins approaches in traffic noise prediction [10]. Support Vector Machines (SVM) have shown a good generalization performance in solving non-linear timeseries problems and have been applied in areas like rock deformation predictions [11]. Gaussian Processes (GP) for regression has also been applied in a variety of fields such as wind power forecasting [12], biological observation and medical health [13, 14].

To conclude, inventory forecasting can be achieved using different data mining approaches. The best approach highly depends on the data being applied to, and methods that are found superior in one context and certain data may not remain so when applied to other datasets and contexts. Since this paper aims to focus more on demonstrating how a data-mining-based approach can be sought by medical stores, probably by non-experts in data mining; and because we want to propose how the estimated forecasting errors can be employed in determining the risks in forecasting, we decided in this study to just emphasize on the importance of experimentally trying different algorithms and techniques rather than committing to a certain method just because it proved better in a similar case study. Therefore, in the design of our experiment, as will be discussed in a later section, we just experimented with common techniques available in the free and open source software Weka, which can be easily accessed by non-experts if they want to try applying our approach to forecast their inventory datasets.

3 Requirements and Proposed Approach

Several meetings and site visits were carried out to identify the problems facing one of the medical stores in Brunei. As mentioned in Sect. 1, stockout of some medicines together with expiry of others have been the main challenges. Most of the medicines take three to four weeks to be available from the time of deciding to place an order. An accurate four weeks forecasting would help in anticipating demand and taking appropriate actions.

3.1 Dataset

Although the real dataset requested from the medical store was not made available due to privacy concerns, a secondary dataset containing daily snapshot products demand was utilized. This secondary dataset resembles the structural records that are currently present in most of medical stores in Brunei, and therefore we expect that the approach taken about this secondary dataset is still applicable in general. The dataset utilized in this study is "Forecasts for Product Demand" and was obtained through Kaggle website[1]. It contains a daily snapshot of products demand from Jan 2011 to Jan 2017, and the structure consists of product code, equivalent to specific medicines in our case, product category, equivalent to the medical category of the drug, product warehouse (indicating the warehouse in which product was demanded) which is equivalent to the sub store in which a certain medicine was demanded and transaction date together with the quantity demanded on that specific date. Although some features of the datasets maybe missing such as prices and number of items remained in stores, our requirements gathering showed that this transactional data represents the basic structure kept by the system utilized in the medical store. Besides, the concerned medical store also kept records for a similar period of time. Therefore, this study and its approach are aimed to be a proof of concept that can be adopted by this medical store, and similar ones in Brunei.

In total, the dataset consists of 1,048,575 dated entries of different products across different warehouses. There are 2,160 different products, encoded to product_code, 4 warehouses, and 33 different product categories.

3.2 Proposed Approach

The proposed approach for this study is depicted in Fig. 1. The approach consists of five stages: (1) Collect the data about previous demand history of all medicines or a subset of concern, and then identify a certain product to forecast; (2) Preprocess the data and records to prepare for applying the selected algorithms, and performing an appropriate lag estimation (can utilize intuitions drawn from data visualizations); (3) Apply a group of time series ML algorithms, which in our study was Linear Regression (LR), Multilayer Perceptron (MLP), Gaussian Process (GP) and Support Vector Machine Regression (SMOreg); (4) Evaluate the performance of obtained

---

[1] https://www.kaggle.com/felixzhao/productdemandforecasting.

models and identify the most accurate one using proper evaluation technique (5) Analyze the risk of using model's accuracy to assist procurement decision making on whether to delay the ordering process or place an early order.

Steps (2) to (4) are quite standard in data mining projects, and they are mostly experimental and could be reiterated until obtaining an acceptable error rate. Our solution utilizes model's accuracy estimation results and provides a risk evaluation mechanism to assist procurement team during ordering process.

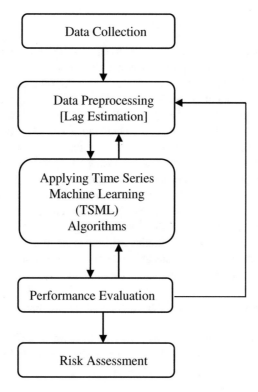

Fig. 1. Proposed approach

3.3   Data Preprocessing and Products Identification

There are different ways that could be used to identify products of interest for forecasting. Below are some criteria that can be considered in selecting these products although our pilot study focused only on one of them:

- Price of the medicine: select expensive medicines and build a forecasting model to help a store optimize on expenses as a main priority.
- Life span of the medicine: focus on forecasting those medicines that have a short life span so that it could help the medical store to order only sufficient quantity of those medication and optimize on expiring products.

- Storage limitations: where focus can be given to relatively large products, or those that need special preservation conditions, etc.
- Frequency: focus on particular medicines that are fast moving; and this was the one utilized in our case study since it was a mentioned factor by the concerned medical store that we analyzed.
- Domain expertise: forecast a selected set of items which are identified by domain experts due to a combination of above reasons, and possibly other factors.

In our study, we identified two different products based on high and low demand frequency, together with one product which was arbitrary selected in the average range. The transactions of these three products were extracted, where product_0979 (low frequently demanded product) had 2,277 transactions, product_1286 (averagely demanded product) had 8,888 transactions, and product_1359 (highly demanded product) had 16,936 transactions. After that, we performed data cleaning to identify incomplete records and detect outliers. In the extracted subset, we found several records having missing values which were then omitted. Also, some of the records contained negative values which might have been wrongly entered. We omitted these values as well since they were not many cases. Finally, we found some cases where the numbers were quite huge, so we identified them as outliers since they were very far from the mean and keeping them would have had a negative impact on prediction models, and thus they were removed as well. Finally, we aggregated our records into a weekly basis which is a common practice as suggested by the medical stores. Hence, a weekly forecasting will enable the medical store to anticipate demands in a number of weeks in advance and take the necessary course of actions.

After aggregating the records, product_1286 had 237 weeks of transactions, product_1359 had 238 weeks, and product_0979 had 257 weeks. Table 1 presents a statistical summary of the products, while Fig. 2 presents a time series plot of them. From the time series plot, we can observe a nonlinear trend for all three products.

Table 1. Statistical summary of the selected products after aggregating the records weekly

	Product_1359 (high demand)	Product_1286 (average demand)	Product_0979 (low demand)
Weeks	238	237	257
Mean	1776651	383745.6	11849
Median	1750000	379900	10800
Min	26000	24400	100
Max	3703000	862500	40200
Std	704221.1	135778.1	8390.718
Variance	4.96E+10	1.843570e+10	7E+07

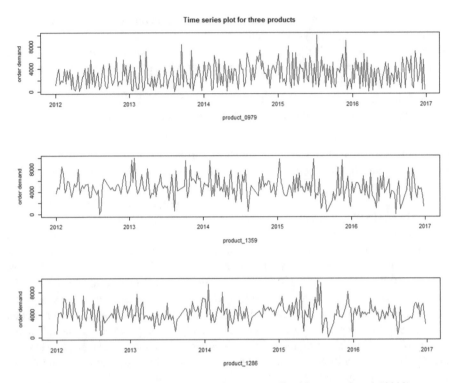

Fig. 2. Time series plot of three products (normalized between 0 and 10000).

3.4   Lag Estimation

One way of getting accurate prediction results is through introducing specific input parameters into the time series data. These inputs can be lagging time variables based on previous records. Choosing appropriate input lag variables is done prior to building the model. Either domain experts' knowledge or visualizing demand trends over a period of time can give insights into the possible lag variables to consider. Increasing the number of lagged variables to go back so long in time will reduce the amount of training data that can be used in learning, while using fewer lagged variables can result in reduced information gained from past trends and thus affecting the predictive power of the models [14]. Furthermore, other approaches have been suggested to improve the predictive power of these variables which include product and power of times and lag variables, and adjusting for variance [15].

3.5   Assessing Risks of Stockouts

For this study, we used Root Mean Squared Error (RMSE) as the performance metric to compare performance of different algorithms, which commonly used in TS evaluation [16]. The following equation shows how to calculate the root mean square error:

$$RMSE = \sqrt{\frac{1}{N}\sum_{i=1}^{N}(y_i - \hat{y}_i)^2} \qquad (1)$$

where y_i indicates the actual/real value of i-th time series point, $\hat{y}_i$ indicates the predicted value, and N represents the number of data points used in evaluating the model.

To address the issue of assessing stockout risk, four mathematical equations have been derived based on predicted demand amount, model's accuracy, expected procurement period, and current medical inventory. The first two equations are:

$$Max = \sum_{j=1}^{P}\hat{y}_j + RMSE_j \qquad (2)$$

$$Min = \sum_{j=1}^{P}\hat{y}_j - RMSE_j \qquad (3)$$

where j here indicates j-week ahead in prediction, $\hat{y}_j$ indicates the predicted value after j weeks, $RMSE_j$ indicates the root mean square error for j-week ahead, and P indicates the required procurement period (in weeks) from placing an order until delivery. Equations (2) and (3) are meant to put upper and lower bounds on expected demand. The *Max* and *Min* values are deduced from model's accuracy per j-week ahead prediction, by interpreting any prediction error as being in the worst-case scenario, for the respective maximum and minimum possible demand. In other words, we do not expect our demand to go above or lower than the values obtained from Eqs. (2) and (3), respectively. These values can be then utilized to create two different risk margins according to the constraints given by the following equations:

$$High\ risk\ if\ (C - Min) < 0 \qquad (4)$$

$$Low\ risk\ if\ (C - Max) < 0 \qquad (5)$$

where C represents the current inventory of the medicine. Equation (3) states that even with the lowest expected demand, the medicine will run out of stock in P weeks' time. Equation (4) means that only with the highest possible demand, the medicine will run out of stock. It should be noted that Eq. (3) should be evaluated first, and then Eq. (4), because *Min* is always less than *Max* (as long as there is error) and thus the later condition implies the former. If none of these constraints is satisfied, it can be assumed that the medicine is in a "safe" demand zone, which can be interpreted as having enough stocks even in the worst-case predicted scenarios for P weeks ahead. These risk margins are assumed to assist procurement staff in estimating the risk situation that will happen based on the type and criticality of the medicine concerned. For example, they might choose to accept being at a "High risk" margin for certain products, but not even at a "Low risk" margin for others. Besides, it is worth noting that the risk intensity could also be measured using the shortage amount calculated from Eqs. (4) and (5), which can further assist the decision makers. For example, bigger expected shortage amounts might be taken more seriously than smaller expected values, which just drop below zero by few items.

It should be clear that improving forecasting accuracy of the models can significantly enhance adaptability of the proposed solution. This is because a largely inaccurate model would trigger purchases more often (bigger High and Low risk margins) than it might be necessary, and hence becoming taken less seriously by procurement decision makers. It is also notable that such risk assessment approach can also be adapted regardless of the utilized time series algorithm, such as the ones specified in this study, since estimated prediction error can be deduced for any of them. Besides, if procurement period differs per medical item, the P value can be changed accordingly.

4   Experimental Results and Discussion

In our study, we applied four weeks ahead forecasting which is equivalent to the procurement period suggested by the medical store. We worked on optimizing the model's parameters to ensure we get the most accurate predictions.

4.1   Utilized Lag Estimation

Because we have used secondary data, determining input lagged variables was done in an experimental approach of trying several weeks back in history and comparing for performance on a 70%–30% training and test splits. For product_0979, we have found that past 15 weeks record, together with similar 15 weeks of the previous year (i.e. weeks 1–15, and 53–68) were the best lag estimations in building all four models. For product_1286, the best lag estimation was obtained by considering the previous 8 weeks together with the previous 3 weeks of past year (i.e. weeks 1–8 and 53–56). Three out of the four models (Linear Regression, Multilayer Perceptron, and Gaussian Process) achieved best performance by using the previous 20 weeks as a lag estimation, while for Support Vector Regression the previous 15 weeks were the best product_1359. We would like to suggest that experimental procedures and parameter tuning examined in our study to be further considered in future applications or studies, since we are aware that these are not necessarily optimal choices.

4.2   Applied Time Series Machine Learning Algorithms

Table 2 presents experimental results for one of the products, product_1286. For each algorithm, the table shows the estimated RMSE values for each of the four weeks ahead predictions.

Our experimental results showed that Gaussian Processes performed better in four-weeks ahead forecasting for two products. Nevertheless, other algorithms may exhibit better performance in different cases depending on the examined time series. Furthermore, we can notice a smaller performance gap between GP and SMOreg. Linear regression as expected did not give an overall good performance which highlights the strength of other machine learning algorithms implemented. Table 3 summarizes the ranking achieved by each method when applied to the selected three products.

Table 2. Model results of product 1286

Product_1286 (1–8, 53–56)				
	Week 1	Week 2	Week 3	Week 4
	RMSE	RMSE	RMSE	RMSE
Linear regression	156394.2	157356.8	148341.5	148571.4
MLP learning rate 0.03	151906.1	149234.4	139969.6	138438.6
GP (Normalized polyKernel = 1.0)	153326.7	154505.8	146083.1	146336.0
SMOreg (Normalized polyKernel = 1.0)	148893.1	147586.9	138213.6	150118.5

Table 3. Ranking of algorithms based on four weeks ahead forecasting

	LR	MLP	GP	SMOreg
product_1359 (High demand)	849518.7 (4th)	829290.4 (3rd)	789551.3 (1st)	809011.2 (2nd)
product_0979 (low demand)	10713.9 (3rd)	12801.6 (4th)	9179.5 (1st)	9619.8 (2nd)
product_1286 (average demand)	148571.4 (3rd)	138438.6 (1st)	146336.0 (2nd)	150118.5 (4th)

5 Conclusion

This study proposed a solution to help the procurement team of a medical store in taking decision of whether to place early order of medicine or to take a risk of delaying this order based on model predictions. The paper also presents an illustration of how data mining can be applied in different medical stores to help forecast their drug inventory with the implementation of different time series machine learning algorithms. Secondary dataset has been utilized in this study, and the exact experimental results and findings will be different when applied to other datasets. These mainly serve for proof-of-concept purpose, since the real datasets from the medical stores are expected to have similar structure. Three different products were selected based on the frequency of demand (high and low demands, together with an arbitrarily chosen product). The study used different machine learning techniques for time series prediction to find the best algorithm that models the timeseries data. Custom lag estimation has significantly improved the models' prediction accuracy. Gaussian Processes has shown an overall better result with a lower RMSE than the rest of the applied methods.

There are some limitations in our study, mainly on the utilization of secondary data, and on the process of determining time lag variables. However, the aim of conducting this study was to demonstrate the potential of applying data mining techniques as an alternative solution to traditional practices employed by medical stores in Brunei Darussalam, to highlight the main challenges, and to propose an approach to be taken by procurement decision makers when interpreting prediction results.

References

1. Matsumoto, M., Ikeda, A.: Examination of demand forecasting by time series analysis for auto parts remanufacturing. J. Remanufacturing **5**(1), 1 (2015)
2. Ghousi, R., Mehrani, S., Momeni, M.: Application of data mining techniques in drug consumption forecasting to help pharmaceutical industry production planning. In: Proceedings of the 2012 International Conference on Industrial Engineering and Operations Management, pp. 1162–1167 (2012)
3. Kumari, A., Prasad, U., Bala, P.K.: Retail forecasting using neural network and data mining technique: a review and reflection categories. Int. J. Emerg. Trends Technol. Comput. Sci. **2** (6), 266–269 (2013)
4. Frost and Sullivan: Independent market research on the global healthcare services (HCS) industry. Indep. Mark. Res. Glob. Healthc. Serv. Ind., 90, June 2012
5. Petry, G.G., Ferreira, T.A.E.: Machine learning strategies for time series forecasting, pp. 2230–2237 (2009)
6. Zadeh, K.N., Sepehri, M.M., Farvaresh, H.: Intelligent sales prediction for pharmaceutical distribution companies: a data mining based approach. Math. Probl. Eng, **2014**, 15 (2014)
7. Anusha, S.L., Alok, S., Shaik, A.: Demand forecasting for the Indian pharmaceutical retail: a case study. J. Supply Chain Manag. Syst. **3**(2), 1–8 (2014)
8. İşlek, İ., Öğüdücü, Ş.G.: A retail demand forecasting model based on data mining techniques. In: 2015 IEEE 24th International Symposium on Industrial Electronics (ISIE), pp. 55–60 (2015)
9. Stefanovic, N., Stefanovic, D., Radenkovic, B.: Applications and Innovations in Intelligent Systems XVI (2008)
10. Garg, N., Sharma, M.K., Parmar, K.S., Soni, K., Singh, R.K., Maji, S.: Comparison of ARIMA and ANN approaches in time-series predictions of traffic noise. Noise Control Eng. J. **64**(4), 522–531 (2016)
11. Li, P., Zhixiang, T., Lili, Y., Deng, K.: Time series prediction of mining subsidence based on a SVM. Min. Sci. Technol. **21**(4), 557–562 (2009)
12. Yan, J., Li, K., Bai, E., Yang, Z., Foley, A.: Time series wind power forecasting based on variant gaussian process and TLBO. Neurocomputing **189**, 135–144 (2016)
13. Andrade-Pacheco, R., Mubangizi, M., Quinn, J., Lawrence, N.: Monitoring short term changes of malaria incidence in Uganda with Gaussian processes. In: CEUR Workshop Proceedings, vol. 1425, pp. 3–9 (2015)
14. Tyralis, H., Papacharalampous, G.: Variable selection in time series forecasting using random forests. Algorithms **10**(4) 114 (2017)
15. Pentaho: Pentaho: Time series analysis and forecasting with weka (2016). https://wiki. pentaho.com/display/DATAMINING/Time+Series+Analysis+and+Forecasting+with+Weka . Accessed 26 Jan 2018
16. López-Yáñez, I., Sheremetov, L., Yáñez-Márquez, C.: A novel associative model for time series data mining. Pattern Recognit. Lett. **41**, 23–33 (2014)

Creative Computing

Learning to Navigate in 3D Virtual Environment Using Q-Learning

Nurulhidayati Haji Mohd Sani[1(✉)], Somnuk Phon-Amnuaisuk[1,2],
Thien Wan Au[1], and Ee Leng Tan[3]

[1] School of Computing and Informatics, Universiti Teknologi Brunei,
Bandar Seri Begawan, Brunei Darussalam
`p20151003@student.utb.edu.bn`, {`somnuk.phonamnuaisuk,twan.au`}`@utb.edu.bn`
[2] Centre for Innovative Engineering, Universiti Teknologi Brunei, Jalan Tungku Link,
Gadong BE1410, Brunei Darussalam
[3] Sesame World Technology, Berakas, Brunei Darussalam
`cyl@sesame-world.com`

Abstract. We implement a self-learning agent capable of navigating through an unknown 3-dimensional (3D) virtual environment. Leverage on Q-learning, an agent gradually learns to navigate the environment through self-exploratory. In contrast to previous works in this area, our state-value function is independent to an agent's position in the environment. Instead, the state-value is designed based on agent-visual perception (i.e. how the environment is perceived through 'raycasting'. This allows a greater flexibility for the agent to reuse its policy in different environment. This approach is also flexible since the Q-table does not grow according to the size of the environment. We investigate the flexibility of knowledge acquisition by examining its performance on different environment setup. The experiment results show that the agent successfully learns to navigate the environment.

Keywords: Navigation · Reinforcement learning · Q-learning

1 Introduction

Reinforcement learning (RL) [14] is a learning paradigm that involves an agent to automatically determines its behavior through repetitive interactions with the environment. Exploration and exploitation are common tactics that RL employs in order to maximize its performance. The agent receives feedback after performing certain action in the environment.

This work is the expansion to previous work by [13], motivated with the aim to build a machine capable of learning to automatically navigate in an unknown environment. We investigate the learning abilities of an agent using RL. We model the agent to navigate automatically in a virtual simulated 3D environment without encoding explicit rules for actions. The agent discovers appropriate sequence of actions using RL. We apply Q-learning [17] to the agent as the learning paradigm and observe its performance in different environmental setup. This

© Springer Nature Switzerland AG 2019
S. Omar et al. (Eds.): CIIS 2018, AISC 888, pp. 191–202, 2019.
https://doi.org/10.1007/978-3-030-03302-6_17

paper investigates: (i) the knowledge acquisition for navigation problem; and (ii) the flexibility of proposed approach for different environment complexity.

The rest of the paper is organized as follows. We review the work related to our investigation in Sect. 2. In Sect. 3, we first formulate the problem and define our environment, and then outline the methods and techniques used for our navigating agent. Section 4 reports the discussions on the experiments and the result to the experiment. Finally, we conclude our findings and propose future work in Sect. 5.

2   Related Works

A self-navigating agent must be able to learn from experiences; avoids obstacles and searches path to the destination. One of the most common learning approaches widely investigated is RL – many in which is used to model an autonomous agent in computer games. An example of this agent is the Fusion Architecture for Learning and Cognition (FALCON) [16], which is a self-organizing neural network agent that implements temporal difference (TD) for behavior modeling in order to play a first-person shooter game. On a similar game, [4] implements Sarsa (λ) algorithm to investigate an autonomous agent that learns to shoot. As interesting as it sounds, these agents are focused on modeling the behavior and leaving the navigation control to random exploration. In contrast, we focus on the navigational aspect of the problem.

Navigation problem have been studied by researchers in various community e.g. robotic swarm [1], automatic vehicle or the taxi-driver [6] and game-like environment [2,3,5]. These studies used a front vision as a fixed viewpoint and from this vision, the agent analyzed the objects before determining its action.

Previous literature approaches navigational problem using well-known methods like genetic algorithm [15] and neural network [8]. Some others proposed new techniques such as context dependent multiagent SARSA (CDM-SARSA) [11]. In contrast to these previous works, which typically use agent's positions as their state-value function, [7,10] approach things differently which is somewhat similar our study. Their agents percept the environment by developing a vision perception of their surroundings and use that as their state values in order to determine the next action.

In our experiment, the agent focuses on navigating its way in the environment, searches for the goal and learns to avoid obstacles along the way using a the visual perception as its states. We also investigate the flexibility of this method for different environmental settings.

3   Learning to Navigate Using Q-Learning

3.1   Problem Formulation

Let E be a virtual 3D environment, populated with m objects O_m. An object may be an agent A, a goal G, or obstacle such as wall W. The agent navigates its way to reach the goal while avoiding obstacles (i.e. walls). In every time step, the agent is able to perform K actions. The agent actions consist of a one-step movement in any of the eight directions (N, NE, E, SE, S, SW, W, NW). An episode ends on either one of the following situations: (i) agent reaches the goal; or (ii) agent wandering the environment for specific maximum steps.

Agent: Agent A is a human model that can perceive its environment through sensors and react accordingly through actions [12]. The agent is assumed to be able to: (i) analyze the environment through perceptors; (ii) acquire knowledge and update its policy through exploration and exploiting its policy; and (iii) perform actions.

3.2   Q-Learning Agent

Q-learning is an algorithm involving the agent to interact with the environment and updating its knowledge based on the actions taken. The agent first percepts the current state at the beginning of each episode, and then decides between exploring and exploiting the actions using ϵ-greedy action-selection policy. Although choosing the most rewarding action to its best knowledge may seemingly to result to a better performance, the agent should also try out random actions because that action probably gives a better outcome. To balance the tradeoff between these exploration and exploitation of policy, a set of rules for action-selection is required. Our agent either takes random action with a probability of ϵ or selects the most rewarding action with the probability of 1— ϵ. In the initial learning process, it is logical for the agent to explore more and exploits more towards the end of the process. Thus, we apply the ϵ-greedy policy with decay where the value of ϵ decreases after each episode. The algorithm for Q-Learning agent is presented in Algorithm 1.

Reinforcement Signals: It is crucial to design a reward mechanism in order for RL to perform successfully because the knowledge of the agent is determined based on it. In our experiment, agent is punished if it hits the wall (−2.0 points); agent is rewarded if it hits the goal (50.0 points), or heading towards goal (0.5 points) if the goal is in sight; and for every step the agent moves, the points is reduced by 0.001. These signals are arbitrary allocated based on the performance after several runs.

3.3   Experimental Setup

There are two major processes involved in this experiment – policy update and policy retrieval (see Fig. 1). In the policy update process, the agent first obtains the state of the environment. Then, one of the two process will take place: selecting a random action; or selecting an action based on its policy. The feedback for taking the selected action for that particular state is received. The process of which the agent chooses the action based on the policy is known as policy

Algorithm 1. Q-Learning

while not max episode **do**
   Initialize $Q(s, a) = 0$
   while not max step or goal not reached **do**
      Observe state s
      Decide between exploration and exploitation using ϵ-greedy policy
      if exploration **then**
         Select random action a
      else
         Select action a with maximum Q-value
      end if
      Perform action a
      Observe next state s' and reward r (if any)
      Update Q:
      $Q(s, a) \leftarrow (1 - \alpha) \cdot Q(s, a) + \alpha \cdot (r + \gamma \cdot \max_{a'} Q(s', a'))$
   end while
end while

retrieval. The action is selected based on the estimated Q-value. The state, the selected action and the feedback are then computed to update its new estimated Q-value for the policy.

The general outline of the environment is shown in Fig. 2(a) and (b). For every new gameplay, exterior walls are positioned stationery to form a squared-shaped room. One goal, one agent and several interior walls are randomly positioned

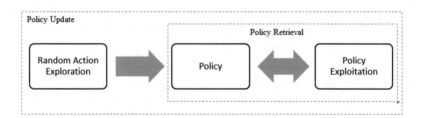

Fig. 1. System architecture for adaptive navigation agent in 3D virtual simulation environment using Q-learning

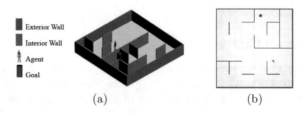

(a)   (b)

Fig. 2. (a) The outline of the environment: an environment, an agent, a goal and walls (obstacles). (b) top view/2D view of the environment

Table 1. Parameter settings for the experiments

Parameter settings	Values		
Environment			
Grid size	6×6	10×10	30×30
Number of walls, W			
Exterior	24	40	64
Interior	15−30	45−90	445−870
Number of agent, A	1		
Number of goal, G	1		
Possible actions, k	$\{N, NE, E, SE, S, SW, W, NW\}$		
Rewards			
Hits a wall	−2.0 point		
Achieve a goal	50.0 point		
Move towards goal	5.0 point		
Every step	−0.001 point		
RL parameters			
Learning rate, α	0.5		
Discount rate, γ	0.99		
ϵ-greedy probability	0.8		

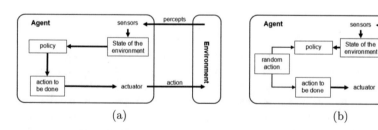

(a)   (b)

Fig. 3. (a) Policy-based Agent: Agent uses state of environment matched with policies to select which action to perform in environment. The agent then updates the new value into the policy; (b) Random-based agent: Agent takes random action to perform in the environment and updates the value to the policy.

within the squared room. The number of walls generated depending on the size of the environment. For this study, we use 6×6, 10×10 and 30×30 which randomly generates random number of walls between 15–30, 45–90 and 445–870 respectively.

An episode ends when a goal is reached or after moving a defined number of steps (see Sect. 4). When an episode ends, the position of the walls and the goal remains and while the agent is relocated randomly. This parameter setup is shown in Table 1.

The process of exploitation, is outlined in Fig. 3(a). This agent is referred to a policy-based agent because it uses the state of the environment and matches its knowledge to select an appropriate action to perform in the environment. The summary of this process is as follows:

(i) Agent percepts the state of the environment using its sensors.
(ii) Agent matches the percepted state to its knowledge, and an action with the highest value is selected.
(iii) The current state of the environment, is paired with the action to be recomputed in order to obtain a new estimated Q-value for the $Q(s,a)$ pair.

An RL agent randomly selects the action upon percepting a state and uses the feedback it receives to estimate the Q-value for that action. This process is also called exploration process as outlined in Fig. 3(b). Below summarizes the process involved in this experiment:

(i) Agent percepts the state of the environment using its sensors.
(ii) Agent selects a random action from all the possible actions.
(iii) Upon receiving feedback (or reward), the current state of the environment, paired with the action are recomputed to obtain a new estimated Q-value for the $Q(s,a)$ pair.

3.4   Experimental Design

We conduct experiments to observe the adaptability of our learning paradigm in different environment settings. In this set of experiments, we conduct different learning environment settings while maintaining the parameter values for Q-learning.

All the experiments are implemented in a box-shaped room as explained in Sect. 4. Each episode is completed either when the agent reached the goal or have reached the defined maximum step. Every experiment is run for 30 times and the results are averaged to reduce noises. Six experiments are carried out to investigate two different comparisons. These experiments comprise of three different grid sizes (i.e. 6×6, 10×10 and 30×30) in two different step limitation (i.e. 36 steps and 300 steps). The settings for all the experiments are summarized in Table 2.

Table 2. Summary of the settings for experiments

Tests	Grid size	# steps	# eps	# run	Description
Exp 1-A	6 × 6	36	300	30	To test if the agent can learn using the parameters
Exp 2-A	10 × 10	36	300	30	To compare performances in bigger grids
Exp 1-B	6 × 6	300	300	30	To compare performances in lesser steps
Exp 2-B	10 × 10	300	300	30	To compare performances in lesser steps
Exp 3-A	30 × 30	36	300	30	To compare performances in bigger grids
Exp 3-B	30 × 30	300	300	30	To compare performances for an increment of steps

Performance Measure. We measure the agent's performance based on the percentages of two parameters: (i) the mean percentage of agent A reaches goal G per number of run (ii) the mean percentage of times the agent hits walls W per run. If the agent could learn from the training, we expect it to navigate itself in the environment to reach the goal while avoiding obstacles after a while. After evaluating the learning ability of our agent using this paradigm, we expect that this paradigm is flexible and adaptable in different grid sizes.

4   Results and Discussion

4.1   Comparison on Different Environment Grid Size

In the first comparison, the adaptation of proposed learning paradigm for the same number of steps in different grid size were investigated. We set the limit to 36-steps for all three grid sizes and the results are shown in Fig. 4. The percentage of hitting a wall is defined as the percentage of total number of times the agent hits the wall over the total number of steps in one episode. The bar indicator in the graph shows the standard deviation. From Fig. 4(a), we can see that the number of wall hits were reduced for all three grid sizes: an average of 0.498 to 0.056 for grid 6×6; an average of 0.430 to 0.035 for grid 10×10; and an average of 0.514 to 0.042 for grid 30×30. Although the percentage of hitting a wall in the chart supports the assumption that the agent learned not to hit walls over time but logically, 36 far too few steps to explore a larger environment on the early episodes of the training resulting to the constant mistakes of the 10×10 and 30×30 in the first 30 episodes.

Another drawback of using less step in a larger environment is that the agent did not have enough time to explore the environment leading to insufficient time to reach the goal (see Fig. 4(b)). Assuming that if the agent initial position is located at one end of the environment and the goal at the opposite end, the approximate average steps would take around 36 steps (6^2) for 6×6, 100 steps (10^2) for 10×10, and 900 steps (30^2) for 30×30. Logically, 36 steps are insufficient for the agent to reach the goal for both 10×10 and 30×30 grids.

We started another experiment in a larger number of steps (i.e. 300 steps). The result in Fig. 5(a) shows that the rate at which agent hits the wall dropped for the first 30 episodes in 10×10 and 30×30 compared to the previous experiment. This is to be expected because the agent was given more time to explore the environment, thus more knowledge was acquired in the early episodes. However, the chart also shows that more mistakes are done in the early episodes for a smaller environment. This is because the agent has limited spaces to roam around and so higher possibility to do the same mistake repeatedly as compared to larger environment. Moreover, when an episode ends, the agent spawned to a random location which also gave the agent a possibility to explore in a new state for a larger environment (30×30).

Finally, Fig. 5(b) shows that more steps are required for the agent to reach the goal in a larger environment. 300 is sufficient number of steps for the agent to reach the goal for 6×6 and 10×10 but still quite insufficient for 30×30. However,

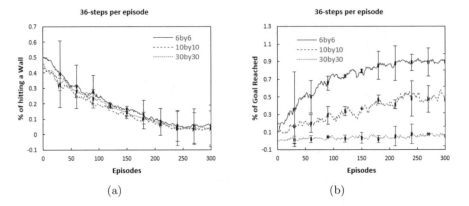

Fig. 4. Results of analysis for different grid sizes with the same number of steps: (a)Averaged percent-age of agent hitting a wall for grid size 6 × 6, 10 × 10 and 30 × 30 in a 36-step limit per episode for 20 runs; and (b)Averaged percentage of goal reached for grid size 6 × 6, 10 × 10 and 30 × 30 in a 36-step limit per episode for 20 runs.

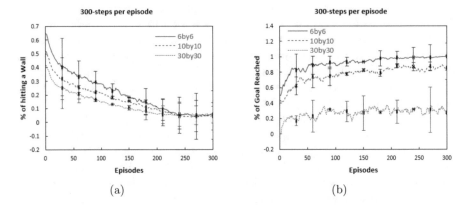

Fig. 5. Results of analysis for different grid sizes with the same number of steps: (a)Averaged percentage of agent hitting a wall for grid size 6 × 6, 10 × 10 and 30 × 30 in a 300-step limit per episode for 20 runs; and (b)Averaged percentage of goal reached for grid size 6 × 6, 10 × 10 and 30 × 30 in a 36-step limit per episode for 20 runs.

there are still possibilities for 30 × 30 to reach the goal with the assumption that the spawn positions for the goal and the agent are near.

4.2   Comparison on Different Number of Steps

In the second experiment, the adaptability of the learning paradigm is investigated for the same grid size but with different number of steps limitation on each episode. There is not much of a difference on the overall performance of 6 × 6 in both 36-steps and 300-steps except for the first 30 episodes as illustrated in Fig. 6(a). Due to a larger step-limit (i.e. 300-steps), it is natural for the agent to commit more mistake in the early stage because more steps to explore means more chances of hitting the wall for the same state. But because of this, the agent also has higher chance of reaching the goal in the early episode (see Fig. 6(b)).

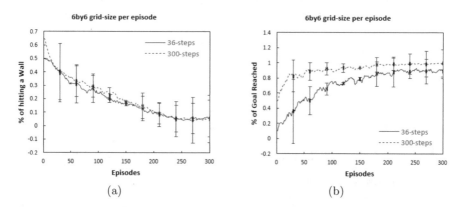

(a)   (b)

Fig. 6. Results of analysis for the same grid sizes with the different number of steps: (a)Averaged percentage of agent hitting a wall for 36-steps and 300-steps limit per episode for 20 runs in a 6 × 6 grid size; and (b)Averaged percentage of goal reached for 36-steps and 300-steps limit per episode for 20 runs in a 6 × 6 grid size.

The performance of agent in a 10 × 10 grid size (see Fig. 7(a)) is somewhat similar to 6 × 6 which the agent learned not to hit walls. However, the result in Fig. 7(b) implies that the agent required more steps to learn to reach the goal.

It is quite impressive that this learning paradigm allows the agent to learn successfully for 30 × 30 grid with only 300 steps (see Fig. 8(a)) even though logically in practice, it would require the agent to move approximately 900 steps to learn the entire grid. However, 36 steps are still incredibly low to learn the entire state. In fact, even 300 steps are insufficient for the agent to search the goal in a larger environment. Figure 8(b) proved that the agent is already having trouble reaching the goal with 300 steps, and worst, the agent is close to not reaching any goal for the with 36-steps. Again, in this case, the agent reaches the goal is only dependent on a chance to be in a close distance during random spawning.

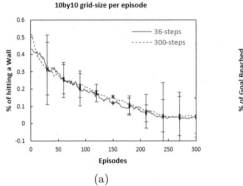 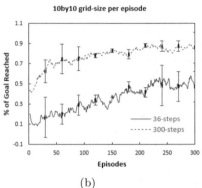

(a)   (b)

Fig. 7. Results of analysis for the same grid sizes with the different number of steps: (a)Averaged percentage of agent hitting a wall for 36-steps and 300-steps limit per episode for 20 runs in a 10 × 10 grid size; and (b) Averaged percentage of goal reached for 36-steps and 300-steps limit per episode for 20 runs in a 10 × 10 grid size.

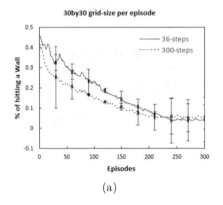 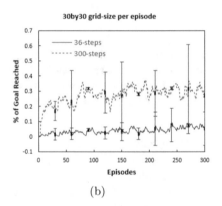

(a)   (b)

Fig. 8. Results of analysis for the same grid sizes with the different number of steps: (a) Averaged percentage of agent hitting a wall for 36-steps and 300-steps limit per episode for 20 runs in a 30 × 30 grid size; and (b) Averaged percentage of goal reached for 36-steps and 300-steps limit per episode for 20 runs in a 30 × 30 grid size.

5   Conclusions and Future Work

In this work, we demonstrated how we create an intelligent RL agent that is able to adapt to a navigation problem in 3D Environment and to learn in real-time through interaction with the environment. Our agent implemented a Q-learning method to learn to navigate flexibly in a variety size environment and in different step limitations. Our agent can learn from scratch and improve its performance and making less mistakes throughout the learning phase.

In our experiment, we analyze and compare two different scenarios. First, we observe its performance on a same set of steps for each episode but operating in different grid sizes. Next, we investigate its performance in a same grid size but different step-limits. Our results show that the current learning paradigm allows the agent to learn on different grid size however the number of steps for each episode, which also determine the time given to the agent to learn, controls these performances. Thus, given a longer learning time, an agent will be able to learn a more complex environment.

Last but not least, we could extend the learning paradigm by either inserting more objects in the environment (as landmarks) as investigated in [9] which may result to an increase in complexity but it may have a possibility to perform better; or by building a model that could represent the map of the environment so that the agent is able to identify the location that has not been chartered. We expect this might be able to help to reduce the redundancy of agent visiting the same spot repeatedly.

Acknowledgement. I would like to thank Sesame World Technologies and Graduate Studies and Research, Universiti Teknologi Brunei for the support and anonymous reviewers for their useful comments and suggestions.

References

1. Atyabi, A., Phon-Amnuaisuk, S., Ho, C.K.: Navigating a robotic swarm in an uncharted 2D landscape. Appl. Soft Comput. **10**(1), 149–169 (2010)
2. Chaplot, D.S., Lample, G., Sathyendra, K.M., Salakhutdinov, R.: Transfer deep reinforcement learning in 3D environments: an empirical study. In: 30th Conference on Neural Information Processing Systems (NIPS 2016) (2016)
3. Dhiman, V., Banerjee, S., Griffin, B., Siskind, J.M., Corso, J.J.: A critical investigation of deep reinforcement learning for navigation. CoRR abs/1802.02274 (2018)
4. Glavin, F.G., Madden, M.G.: Adaptive shooting for bots in first person shooter games using reinforcement learning. IEEE Trans. Comput. Intell. AI Games **7**(2), 180–192 (2015)
5. Hussein, A., Elyan, E., Gaber, M.M., Jayne, C.: Deep imitation learning for 3D navigation tasks. Neural Comput. Appl. **29**(7), 389–404 (2018)
6. Isele, D., Cosgun, A., Subramanian, K., Fujimura, K.: Navigating intersections with autonomous vehicles using deep reinforcement learning. CoRR abs/1705.01196 (2017)
7. Jaafar, J., McKenzie, E.: Autonomous virtual agent navigation in virtual environments. World Acad. Sci. Eng. Technol. **37**, 594–601 (2010)
8. Lozano, M., Vilaplana, J.M.: A neural approach to an attentive navigation for 3D intelligent virtual agents. In: Proceeding of the IEEE International Conference on Systems, Man and Cybernetics, vol. 6 (2002)
9. Mirowski, P., Pascanu, R., Viola, F., Soyer, H., Ballard, A.J., Banino, A., Denil, M., Goroshin, R., Sifre, L., Kavukcuoglu, K., Kumaran, D., Hadsell, R.: Learning to navigate in complex environments. CoRR abs/1611.03673 (2016)
10. Muhammad, J., Bucak, I.O.: An improved Q-learning algorithm for an autonomous mobile robot navigation problem. In: Proceeding of the 2013 The International Conference on Technological Advances in Electrical, Electronics and Computer Engineering (TAEECE), pp. 239–243 (2013)

11. Phon-Amnuaisuk, S.: Learning cooperative behaviors in multiagent reinforcement learning. In: Proceeding of the Neural Information Processing, pp. 570–579. Springer, Heidelberg (2009)
12. Russell, S., Norvig, P.: Artificial Intelligence: A Modern Approach. Pearson (2010)
13. Sani, N.H.M., Phon-Amnuaisuk, S., Au, T.W., Tan, E.L.: Learning to navigate in a 3D environment. In: Proceeding of the Multi-disciplinary Trends in Artificial Intelligence, pp. 271–278. Springer (2016)
14. Sutton, R.S., Barto, A.G.: Reinforcement Learning: An Introduction. MIT Press, Cambridge (2012)
15. Velagic, J., Lacevic, B., Peruničić-Drazenovic, B.: A 3-level autonomous mobile robot navigation system designed by using reasoning/search approaches. Robot. Auton. Syst. **54**(12), 989–1004 (2006)
16. Wang, D., Tan, A.H.: Creating autonomous adaptive agents in a real-time first-person shooter computer game. IEEE Trans. Comput. Intell. AI Games **7**(2), 123–138 (2015)
17. Watkins, C.J., Dayan, P.: Q-learning. Mach. Learn. **8**, 279–292 (1992)

Pixel-Based LSTM Generative Model

Somnuk Phon-Amnuaisuk[1,2,3(✉)], Noor Deenina Hj Mohd Salleh[3],
and Siew-Leing Woo[3]

[1] Media Informatics Special Interest Group, CIE, Universiti Teknologi Brunei,
Bandar Seri Begawan, Brunei Darussalam
`somnuk.phonamnuaisuk@utb.edu.bn`
[2] Centre for Innovative Engineering, Universiti Teknologi Brunei,
Bandar Seri Begawan, Brunei Darussalam
[3] School of Computing and Informatics, Universiti Teknologi Brunei,
Bandar Seri Begawan, Brunei Darussalam
`deenina.salleh@utb.edu.bn, m20083001@student.utb.edu.bn`

Abstract. Applying computational intelligence techniques to create
generative models of digits or alphabets has received somewhat little
attention as compared to classification task. It is also more challenging
to create a generative model that could successfully capture styles and
detailed characteristics of symbols. In this paper, we describe the application of the Long Short-Term Memory (LSTM) model trained using a
supervised learning approach for generating a variety of the letter **A**.
LSTM is a recurrent neural network with a strong salient feature in its
ability to handle long range dependencies, hence, it is a popular choice
for building intelligent applications for speech recognition, conversation
agent and other problems in time series domains. To formulate the problem as a generative task, all the pixels in a 2D image representing an
alphabet (i.e., the letter **A** in this study) are flattened into a long vector
to train the LSTM model. We have shown that LSTM has successfully
learned to generate new letters **A** showing many coherent stylistic features with the original letters from the training sets.

Keywords: Pixel-based generative model · Long short-term memory
Recurrent neural network

1 Introduction

Human beings have their own object of reality constructed from each of their
own individual experiences. These experiences are based on how much attention
they have given to their senses. Their perceptions are represented as representa-
tional features which are retained in their Short-Term (ST) and Long-Term (LT)
memory [1]. If you are asked to draw a representation of the following objects:
the moon, the sun, a star, a heart, a lightning bolt and a bat. It is very likely
that you will draw a crescent shape to represent the moon, a circle to represent
the sun and a five-pointed shape to represent a star, and so on (see Fig. 1). We

© Springer Nature Switzerland AG 2019
S. Omar et al. (Eds.): CIIS 2018, AISC 888, pp. 203–212, 2019.
https://doi.org/10.1007/978-3-030-03302-6_18

argue that this is because the crescent shape, the spherical shape and the five pointed shape have been established as representative iconic figures of the moon, the sun and the stars respectively.

Folk psychology supports the concept of mental imagery where perceptual experiences (e.g., visual, motor, and haptic imagery) are represented mentally. In the suggested experiment above, one reasonable explanation is that our mental representations associate shapes to describe those objects. Human associates the crescent shape with the moon since we have seen the moon in its crescent shape more often than in a round shape. There is only one full moon in thirty days. Following this line of thought, objects with a complex shape can be perceived as constructed from basic *geons* (aka geometrical ion) as postulated in the *geon* theory by Biedermann [2]. Although this line of thought does not clearly explain how patterns of activations on the retina could result in mental imagery of geometrical representations, there is a clear logical link between the low level activation of retina cells to the basic shapes and the high level representation of visual symbols. Advances in neural computing and cognitive science show various pieces of evidence supporting this concept. For example, neuronal cells show conditional responses to visual stimulus patterns [3]. This suggests that primitive shapes are activated by different functional units and they are hierarchically combined to form a more complex structure. A recent study on convolutional neural network (CNN) by [4,5] also shows that the convolutional layers of CNN appear to learn hierarchical representation of visual stimuli.

Fig. 1. Icons map perceptions from senses to mental imagery. First row: Icons of various visual observations. Second row: Complex objects can be constructed from basic geometrical shapes [2].

Motivated by the idea that higher level structure could be constructed from low level primitives, we explore a computer based generative model, in particular, a pixel-based generative approach using the LSTM model. In this work, the focus is not on the representation of geometrical shapes but on the fact that the high level representation is constructed from the low-level pixel information. We train the LSTM recurrent neural network using the typeface **A** obtained from the notMNIST dataset[1]. Our implementation shows that the LSTM model can

---
[1] yaroslavvb.blogspot.com/2011/09/notmnist-dataset.html.

successfully learn the representation of a typeface **A** and the model can be used to generate new instances of **A**.

The rest of the paper is organized into the following sections: Section 2 discusses the pixel-based generative process and gives details of the LSTM technique; Section 3 provides the experimental design and results; Section 4 provides qualitative analysis as well as a critical discussion of; and finally, the conclusion and further research are presented in Sect. 5.

2   Pixel-Based Generative Process

In this implementation, the value of each pixel is simplified to either 1 or 0 (white or black)[2]. Let $<x_1, ..., x_n>$ be a sequence of pixels created by flattening an image $X^{r \times c}$, where $r \times c = n$. Let a pair (I, P) be a training sample where I is a sequence of input pixels and P is the predicted pixel value of the next pixel. A training data for a predictive model m can be prepared from a pixel-subsequence taken from X, say use three pixels to predict the fourth pixel, e.g., $<x_1, x_2, x_3> \mapsto x_4'$; $<x_2, x_3, x_4> \mapsto x_5'$, and so on.

If we construct a predictive model of a given graphical symbol, then the model can be employed to generate a new instance of the symbol. There are many choices of computational models for a prediction/classification task. A probability model, a Feedforward Neural Network (FFNN), or a Recurrent Neural Network (RNN) may be employed. In this work, we have chosen RNN for its salient features in handling sequential input, in particular, a variant of RNN known as the Long Short-Term memory (LSTM) model [6].

2.1   Long Short-Term Memory (LSTM)

The RNN network is traditionally designed to handle delayed input signal e.g., $Y_t = \sigma(WX + b)$ where σ is a transfer function and $X = X_t | Y_{t-1}$ (i.e., X is a concatenation of X_t and Y_{t-1}). Different durations of delay time signify different amount of memory the network takes into account when determining the output.

LSTM was proposed in 1997 by Hochreiter and Schmidhuber, it is a variant of RNN. LSTM introduces new concepts of a memory cell, an input gate i, an output gate o and a forget gate f. The LSTM state represents the memory cell and the gates control the extent of information that flows into, is retained and flow out of the memory cell. The amount of information that flows between the gates of LSTM units is controlled by the weights of the LSTM network. The memory cell concept allows LSTM to effectively handle the issue of long-range dependency betters than the traditional RNN.

Figure 2 (top row) illustrates (a) a single LSTM unit and its roll-out representation; (b) LSTM units can be stacked up to form a multi-layer unit. At time t, an input X is the concatenation of $X_t | H_{t-1}$ where X_t is the input at time t and H_{t-1} is the delayed input from the previous time step. Figure 2 (bottom

---

[2] Hence, our predictive model can also be thought of as a binary classification model.

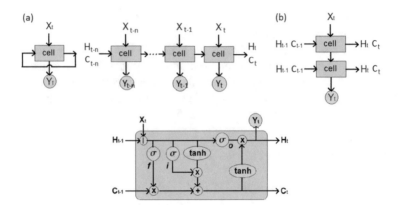

Fig. 2. Top row (a): A single LSTM cell with a recurrent input is the abstraction of a sequence of LSTM cell across time; (b) Many LSTM cells may be stacked to create a deep structure. Bottom row: Details inside the LSTM cell is illustrated.

row) illustrates the flow of information: $H_{t-1} \Rightarrow H_t$; $C_{t-1} \Rightarrow C_t$ and $X_t \Rightarrow Y_t$ in a LSTM unit. More detailed explanation of the LSTM cell unit is expressed in the following equations:

$$f_t = \sigma(W_f[X_t, H_{t-1}] + b_f) \tag{1}$$

$$i_t = \sigma(W_i[X_t, H_{t-1}] + b_i) \tag{2}$$

$$o_t = \sigma(W_o[X_t, H_{t-1}] + b_o) \tag{3}$$

$$C'_t = tanh(W_c[X_t, H_{t-1}] + b_c) \tag{4}$$

$$C_t = (f_t \otimes C_{t-1}) \oplus (i_t \otimes C'_t) \tag{5}$$

$$H_t = o_t \otimes tanh(C_t) \tag{6}$$

$$Y_t = softmax(W H_t + b) \tag{7}$$

where the initial values are $C_0 = 0$ and $H_0 = 0$ and the operator $\otimes$ denotes the Hadamard, element-wise product. Matrices W_g denote the weights of the input and the recurrent input, where g can either be the input gate i, output gate o, the forget gate f or the memory cell c, depending on the activation being calculated. The function σ denotes the sigmoid function, *tanh* denote the hyperbolic tangent function and *softmax* is the normalized exponential function.

3   Constructing a Generative Model from LSTM

To construct a generative model using deep learning approach, a large collection of training example is needed. This work employs the notMNIST dataset as the training data. The dataset contains ten classes of letters **A, B, ..., J**. Each letter is represented as a gray scale image with the size of 28×28 pixels^2.

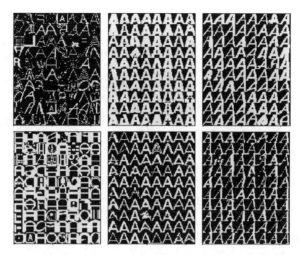

Fig. 3. Examples of the letter **A** randomly picked from the six clusters. These images will be referred to as image set one, two, ..., six (reading from left to right and from top to bottom).

Here, the letter **A** from the dataset will be modelled. Upon inspection, there are over 50,000 examples of **A** in the notMNIST dataset. A variety of typeface are observed, hence, it is decided that these samples should be grouped together according to their similarity first and then separate models should be trained from these groups. The grouping is ad-hoc using the *k-means* clustering method [7]. The similarity measure is based on the similarity of pixels among images which loosely capture the typeface style in its metric. Figure 3 shows examples of **A**s randomly chosen from six different clusters. Six different predictive models are constructed, one for each cluster group.

Generated Fonts from Pixel-Based LSTM Model: There are many variants of LSTM in the literature. In this experiment, we employ the *Tensorflow* implementation from `tf.contrib.rnn.BasicLSTMCell` follows [8]. Figure 4 shows a high level conceptual model of our setup. Each LSTM cell has one input $X_t \in \mathcal{R}^d$ which represents one pixel, 128 hidden units $H_t \in \mathcal{R}^h$ and one output $Y_t \in \mathcal{R}^d$. Our model is trained with three layers of LSTM with a sequence length of 400 cells. Hence, the LSTM learns to map the training data pair $T \mapsto V$:

$$<x_{n-400}, x_{n-399}, ..., x_n> \mapsto <x_{n-399}, x_{n-398}, ..., x_{n+1}>$$

After the model has successfully learned the letter **A**, the model can be employed to generate an unseen letter pixel by pixel. An image of size 28×28 would require 784 generation steps. Figure 5 shows 36 instances of the letter **A**, six instances per group. Comparing these generated **A** to the training examples in Fig. 3, we are convinced that the model has successfully learnt the model of **A**. However, it will be a challenge to find a way to objectively quantify how well the model has learnt.

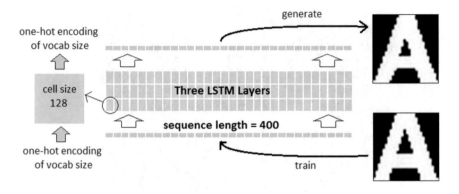

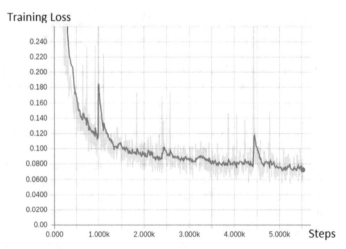

Fig. 4. Top pane: the architecture of our generative model. Bottom pane: A typical training loss observed for all groups of **A**. This shows that the model has successfully learned to generate the symbol **A**.

4   Qualitative Analysis and Discussion

Due to the subjective nature of the domain, we design an experiment attempting to investigate whether humans are able to differentiate between the notMNIST dataset and the generated data.

Materials: One hundred and forty four examples of the letter **A** were randomly selected from the six clusters of notMNIST dataset. These sets are known as the training-examples sets. Another 144 examples are generated using the six trained LSTM models. These sets is known as the generated-examples sets. Since the grouping by k-means was loosely based on the correlation of pixels and the visual weight of the letters, such groupings would help to create a pattern for

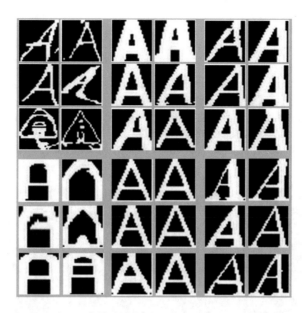

Fig. 5. Thirty-six instances of the letter **A** from the training-examples sets and thirty-six instances of the letter **A** generated from the six models are displayed side by side. Six representative examples for each group are shown here. These instances shows a clear similarity between the instances from both groups.

the participants to evaluate with where in total, there were six sets in which each of them had 24 examples. The characteristics of these sets are summarized below.

Image-set one were grouped based on light fonts, where the weight of the typeface letters appears to be thinner; image-set two are bold letters which has a thicker weight on the letter; image-set three are bold italic letters which are thicker and slanted; image-set four are letters which are extensively bold where the letters are thicker and wider in nature; image-set five are Roman type letters where the letters are considered to have the standard weight of a typeface and are in an upright form; and image-set six grouped italic fonts where the letters have a standard weight and are slanted. Such groupings would help to create a pattern for the participants to evaluate on since it provides a consistent shape (see [9], p. 83). These comparisons are distinguishable due to the similarity of the letter's structure.

The generated-examples sets were then compared to the training-examples sets side by side in different orders for each comparison. Setting the context by having the training-examples set and generated-examples set together is done to create a sense of visual coding among the audience (see [10], p. 274); this in a way, ensures the audience to unconsciously decode the visual chaos being presented to them, forcing them to understand the content and how to evaluate the information as well as maintain a stable perceptual experience. Having such a cognitive engagement with visual stimuli provides a more satisfying experience to their participation (see [11], p. 125).

Data Collection: The study was conducted as an online survey on Google Sites. Participants were asked to fill in some information regarding themselves: name, age and gender; as well as to select the group of images that seemed to be the generated dataset based on their perception for each of the six questions (see six image-sets in Fig. 6). The web link was then disseminated via WhatsApp.

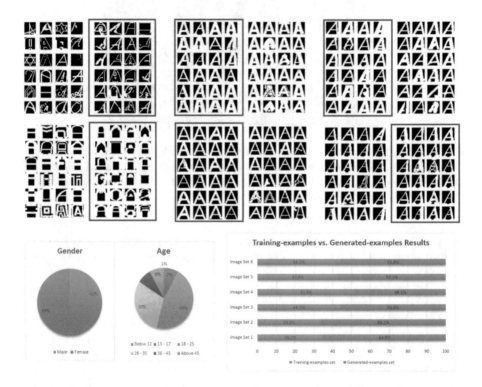

Fig. 6. Rows 1 and 2: Six image sets of training-examples and generated-examples are put side by side. Boxes are marked on the generated-examples sets. These instances show a clear similarity between the instances from training-examples and generated-examples. Row 3: summary of the participants' profiles and the outcome of the survey collected from 77 participants. Participants correctly identify the generative-examples for all image-sets, except image-set four.

4.1   Results and Discussion

According to the status report, a total of 77 participants volunteered to answer the survey. Out of those participants, 58% were females. The age group varied overall: 1% of the response was below 12 years old, 7% aged between 13 to

17 years old, 46% were between 18 to 25 years old, 30% of the participants were between 26 to 35 years old, 8% ranged between 36 to 45 years of age and another 8% were aged 45 and above. Their evaluations of the six sets of images were recorded and summarized in Fig. 6.

Upon evaluation, it was found that five out of the six compared Image-sets, participants were able to correctly identify the generated-examples set of letter **A**, for image-set one it was 64.9%, 66.2% for image-set two, 55.8% for image-set three, 57.1% for image-set five and 55.8% for image-set six. However, 51.9% of them wrongly chose the training dataset instead of the generated dataset for image-set four. This was a surprise outcome to us as we had not expected an easy identification since the image results of the generated **A** contain minimal to no difference from the training-examples sets, therefore this should make it indistinguishable to the participants.

With a careful visual inspection of instances from training-examples sets, generated-examples sets, and insights from interviewing the participants, we hypothesize that, in the context of our experiment, participants tend to perceive that computer generated letters are imperfect i.e., with jagged edge or some seemingly wrong pixels. This may be due to the fact that we are often exposed to the perfection of **A**s that we see in our everyday lives in the paper or other media. Hence, an imperfect letter **A** is not a familiar sight. There is evidence from literature that the brain forms concepts based on the consensus of neural activities during visual experiences [12], hence familiarity could influence perceptions. This may explain why participants inclined to perceive the perfect letter **A**s to be the original dataset and tend to identify imperfection as computer generated fonts in the context of this experiment. This hypothesis deserves a careful investigation in our future work.

Relationships to Related Works: Generative models can be coded as a compact rewrite rules such as those in the L system [13], or can be a long procedural sequence such as in the examples given by AARON[3], one of the pioneer works that paints images using program instructions. The procedural approach is effective for describing a top down generative approach. Later works explore the approach in many facets, cellular automata [14], procedural painting [15], evolving 3D models [16] and the recent *Generative Adversarial Network* (GAN) [17].

5   Conclusion

We have constructed a pixel-based LSTM generative model capable of generating the letter **A** in the style of the original training examples e.g., light font, roman type, italic, bold face, etc. This work shows that LSTM can be implemented as a generative model. In this implementation, the pixels in 2D images are preprocessed and converted as a time series vector before segmented as Input-Predicted

---

[3] https://en.wikipedia.org/wiki/AARON.

(I, P) pairs to train the LSTM. In our future work, we wish to expand the repertoire of the models to handle complex symbols and look deeper into the issue of humans pre-assumption and perception of computer generated symbols.

Acknowledgments. We wish to thank anonymous reviewers for their comments that have helped improve this paper. We would like to thank the GSR office for their partial financial support given to this research.

References

1. Norman, D.A.: The Design of Everyday Things. MIT Press, London (2013)
2. Biederman, I.: Recognition-by-components: a theory of human image understanding. Psychol. Rev. **94**(2), 115–147 (1987)
3. Hubel, D.H., Wiesel, T.N.: Receptive fields of single neurones in cat's striate cortex. Phisiology **148**(3), 574–591 (1959)
4. Zeiler, M.D., Fergus, B.: Visualizing and understanding convolutional networks. In: Proceedings of the European Conference on Computer Vision (ECCV 2014), pp. 818–833 (2013)
5. LeCun, Y., Bengio, Y., Hinton, G.: Deep learning. Nature **521**, 436–444 (2015)
6. Hochreiter, S., Schmidhuber, J.: Long short-term memory. Neural Comput. **9**(8), 1735–1780 (1997)
7. Forgy, E.W.: Cluster analysis of multivariate data: efficiency versus interpretability of classifications. Biometrics **21**, 768–769 (1965)
8. Zaremba, W., Sutskever, I., Vinyals, O.: Recurrent neural network regularization. http://arxiv.org/abs/1409.2329 (2015)
9. Arnheim, R.: Art and Visual Perception: A Psychology of the Creative Eye. University of California Press, London (1974)
10. Santoro, S.W.: Guide to Graphic Design. Pearson (2014)
11. Costello, V., Youngblood, S.A., Youngblood, N.E: Multimedia Foundations: Core Concepts for Digital Design, 2 edn. Focal Press (2013)
12. Pollen, D.A.: On the neural correlates of visual perception. Cereb. Cortex **9**(1), 4–19 (1999)
13. Prucinkiewicz, P., Lindenmayer, A.: The Algorithmic Beauty of Plants. Springers, New York (1996)
14. Wolfram, S.: Cellular automata as models of complexity. Nature **331**(4), 419–424 (1984)
15. Phon-Amnuaisuk, S., Panjapornpon, J.: Controlling generative processes of generative art. In: Proceedings of the International Neural Network Society Winter Conference (INNS-WC 2012). Procedia Computer Science, vol. 13, pp. 43–52 (2012)
16. Ariffin, M.K., Hadi, S., Phon-Amnuaisuk, S.: Evolving 3D models using interactive genetic algorithms and L-systems. In: Proceedings of the 11th International Workshop on Multi-disciplinary Trends in Artificial Intelligence (MIWAI 2017), pp. 485–493 (2017)
17. Goodfellow, I., Pouget-Abadie, J., Mehdi, M., Bing, X., Warde-Farley, D., Ozair, S., Courville, A., Bengio, J.: Generative adversarial networks. http://arxiv.org/abs/1406.2661 (2014)

Implementation of Low-Cost 3D-Printed Prosthetic Hand and Tasks-Based Control Analysis

Saiful Omar[✉], Asem Kasem, Azhan Ahmad, Seri Rahayu Ya'akub,
Safwan Ahman, and Esa Yunus

Universiti Teknologi Brunei, Bandar Seri Begawan, Brunei Darussalam
{saiful.omar, asem.kasem, azhan.ahmad,
serirahayu.yaakub}@utb.edu.bn,
safwanbrunei@gmail.com, esa.yunus@mindef.gov.bn

Abstract. A functional prosthetic hand can cost up to £10,000 which limits its access to many amputees. With advancement of technology, one of the possibilities to overcome this issue lays in the use of 3D-printing. A 3D-printer can reduce the production cost significantly, to less than £400, for models that can achieve basic functionalities. There have been several developments of 3D-printed prosthetic hands and arms, and some of them have been made open source. This paper presents a work in progress of implementing a 3D-printed prosthetic hand based on an open source model, describes some of the important issues and challenges faced, and carries out a tasks-based control analysis for some activities of daily living; namely those that depend on power, tip, lateral, and spherical grasps.

Keywords: 3D-printing · Prosthetic hand · Bionic arm · Amputees

1 Introduction

A prosthetic is an artificial substitution for a missing limb. The limb can be lost through an accident, birth defect or an illness. Prostheses are meant to be functional such as to help in the daily life activities of an amputee. However current prostheses have limited functionality and can be expensive to own. In many places around the world, it is financially challenging for public healthcare systems to provide amputee patients with commercial prosthetics. This makes it inaccessible to some amputees who cannot afford it. Moreover, amputees tend to abandon using heavy prostheses [4]. Because of the cost and weight issues, many amputees prefer to use cosmetic prostheses, which however provide poor functionality. With the advances in (three dimensional) 3D-printing, it is now possible to 3d-print light-weight prosthesis at a low-cost, making it a more viable option for many amputees, especially in places affected by wars where the number of amputees is high and accessibility to professional healthcare is limited.

Some open source hardware projects dedicated to prostheses are available online, such as OpenBionics [8] and HACKberry exii [6]. These projects make hardware models available to the public, which improves accessibility by researchers and

© Springer Nature Switzerland AG 2019
S. Omar et al. (Eds.): CIIS 2018, AISC 888, pp. 213–223, 2019.
https://doi.org/10.1007/978-3-030-03302-6_19

interested people for experimentation and conducting further development. Therefore, the purpose of our research is to develop a lightweight and low-cost (3D)-printed prosthetic bionic hand, with suitable sensory and control optimization to obtain acceptable usability in performing basic activities of daily living.

To reach the aim, the following objectives are targeted:

- Produce a low-cost 3D-printed prosthetic hand.
- Utilize arm-attached infrared distance sensor to control grasps of prosthetic hand using different fingers movements.
- Calculate total cost of prosthetic.
- Fine-tune the sensory and hand control to optimize performance of certain tasks.
- Study the possibility of introducing other interfaces to control the hand, such as EMG sensor, or multiple infrared sensors, while keeping low production cost.

Section 2 of this paper discusses further on open source 3D-printing prosthetic project used in this study, and reviews the sensor used and discusses sensor placement that enable better control. Section 3 shows our methodology and how we have tested the arm for functionality to identify what needs improvement. Section 4 evaluates the results of the testing done in earlier section and shows the total cost of producing the prosthetic arm. The last section of this paper concludes by summarizing the evaluation presented and points for improvements to be made in the future.

2   Background on 3D-Printed Prosthetics

Over the last 5 years, significant development has occurred in 3D-printing technologies. To utilize 3D-printing for prostheses, it is important to consider the following advantages and disadvantages:

2.1   Advantages

3D-printing has several advantages compared with other manufacturing techniques, it makes it possible to produce products out of one part; therefore, no assembly is required [1], vast design freedom; therefore, highly complex geometries can be made [2], designs can easily be personalized and customized; parts can be produced cheaply and quickly from idea to end product with rapid design improvements [3].

2.2   Disadvantages

3D-printing also has disadvantages compared to other manufacturing technique such as being hard to predict the mechanical properties. The resulting strength of a part is highly dependent on the fabrication method, and various parameters can be selected depending on the printing orientation; accuracy is highly affected by material shrinkage; different machine parameters and errors induced by the CAD/CAM [5] software as well as post processing [3]; size of an object is limited by the size of the printer, large objects cannot be basic 3D-printing technology, 3D-printers can work with a limited amount of materials compared with conventional manufacturing that can work with nearly any material [3].

2.3 Producing the Prosthetic

For this project, we are going to be concentrating on trans-radial amputees. Trans-radial amputees refer to the partial amputation of arm below the elbow. There have been few developments of open source prostheses projects for this type of amputation, which are available to download and 3D-print at home. In this study, we have used the HACKberry exii [6] design. The hand uses an Arduino [7] to control the movement of fingers and parts [16], and it has different modes that can be switched using programmable buttons. This makes the hand customizable and enables different types of grasps, such as lateral, power and precision grasp [3]. The stereolithography (STL) file is available online on Github, and we have 3d-printed it using polylactic acid (PLA) filament. We have assembled the hand using the provided online instructions and utilized the designer's controller and motors. The assembly of the hand requires us to solder most of the parts to the given PCB board including the sensor and servo motors. The hand uses 3 servo motors; 1 large servo to control the index finger and the other 2 servos to control the thumb and the other 3 finger movements.

To control the prosthetic, sensors must be used to classify signals from amputee's remaining limb into controlling commands. Classification accuracy is important so that the prosthetic can move according to what the amputee user intends to. The main sensors that could be considered for receiving commanding signals are Electromyography (EMG), Infra-Red (IR) and Electroencephalography (EEG) sensors. In this study we have used the provided IR sensor, mainly for cost reasons, and will be discussing the possibility to include an EMG sensor in the last section.

2.4 The IR Sensor

The sensor that is used for HACKberry is an IR sensor that detects the distance between the skin surface and the sensor. When it is placed at the appropriate position at the forearm, it should be able to detect the flexion of the muscles beneath it. If a flexion happens, it signifies that the prosthetic hand should close. The clear advantage of this sensor is that it does not produce noise such as EMG and is a cheaper alternative. However, the disadvantage is that it is difficult to distinguish the signals to move separate fingers as it used to close and open the hand. IR sensor placement and signal classification.

As mentioned in earlier, the IR sensor is able to detect the flexion and extension of muscle by measuring the distance between skin surface and the sensor itself. For a flexion of the hand to occur, it uses four muscles; Brachioradialis, Flexor Carpi Radialis, Flexor Carpi Ularis and Palmaris. However, to get the best reading from the sensor, the IR sensor should be placed on the Brachioradialis near the elbow. This is where the muscle is the biggest on the radial side of the arm.

It is important to flex the muscle during the calibration stage so that it can get the correct minimum and maximum sensor values. The Arduino then processes the values to move the prosthetic hand. Maximum value is the value the sensor receives when the muscle is fully relaxed. Minimum value is the value the sensor receives when the muscle is fully contract. Both maximum and minimum value can be seen in Fig. 1 when flexion and relaxation occurred. The value in between refers to when the muscle is nether fully contract or relaxed.

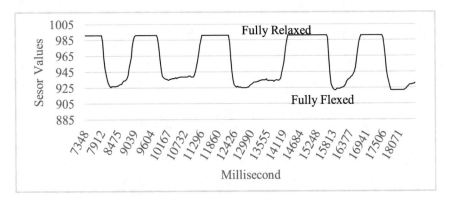

Fig. 1. Muscle flexion signal over time

From the maximum and minimum value, it is then mapped to the position of the index fingers and the other 3 fingers by mapping the value of sensor to the position of the fingers.

3   Methodology

This is the proposed methodology to test the HACKberry prosthetic hand. The methodology below aims to identify the changes that can be done to the HACkberry arm to fully optimize it. We used five people who are not amputees to do the testing below. For each volunteering person:

1. *Calibration test:* Classify best tightness of the IR sensor so it can best receive the signal during calibration.
2. *Grasping test:* Test all volunteer if they are able to perform different type of grasp.
3. *ADL Test:* Perform a variety of tests of Activity of Daily Living.
4. *Evaluation:* Evaluate the results of the test and identify what needs to be changed on the HACKberry arm from the conducted tests.

Before going into the details of calibration, grasping and ADL tests, we first explain the concept of adaptive grasp.

3.1   Adaptive Grasp for a Functional Hand Prosthetic

For a prosthetic hand to be functional, the grasp needs to be adaptive. An adaptive grasp has the ability to hold an object of any shape using the hand, where the force is distributed to ensure that some fingers can still apply a force when the other fingers are halted by an object [10]. As seen in Fig. 2, according to the Southampton Hand Assessment Procedure (SHAP) [11], there are six grasp types; (1) Tip, (2) Lateral, (3) Tripod, (4) Spherical, (5) Power, and (6) Extension.

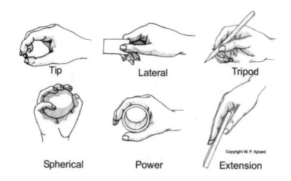

Fig. 2. Different grasps classification according to SHAP [11]

Ideally, all hands should be able to perform all types of grasps. However due to the limited functionality of the current prosthetic hand, they are only able to perform some of the grasps as will be demonstrated by the grasping tests.

3.2 Calibration Test

The calibration test determines the best tightness for the sensor to be strapped on a person's limb. This is done by using spring scale measurement to measure how tight the sensor is strapped. We decided to classify the tightness into 3 categories; Class A (Not tight), Class B (Tight) and Class C (Very Tight). By using a spring scale measurement, we can divide the tightness into those three categories. The end of the sensor strap is tied to the spring scale for this purpose. Before fully strapping the sensor on the forearm, the value of the spring measure is taken based on how tight the volunteer feels it is. The result of the best tightness from this test will then be used for the Grasping test and the ADL test.

Volunteer comfort level correlates with the tightness of the strap which was measured using a spring scale in Newtons (N). This is the average value of the 5 volunteers:

$$Class\,A = 4.01N \quad Class\,B = 11.28N \quad Class\,C = 21.57N$$

After knowing the scale value for the tightness for each category, we also tested which of those classification receives signal from the IR sensor best during calibration. This is done by measuring the sensor value received from the IR signal and see if the HACKberry hand closes and open properly. From the signal received we are also taking note in the maximum and minimum value and see whether in a full flexion the signal received is able to reach the value and properly close the hand. This was further

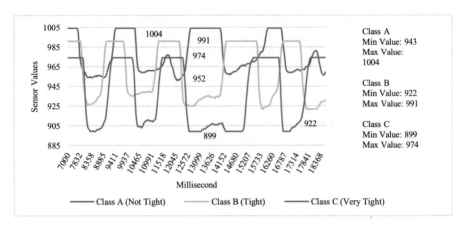

Fig. 3. Sensor value from volunteer 1 as a line graph for 3 tightness classification.

explained in Sect. 2.4. As seen in Fig. 3, Class A is not able to reach the minimum value which is 943 and they are only able reach 952 at full flexion whilst also still struggling.

This can cause a problem as it means the prosthetic hand will be not be able to properly close and flexing too hard to try and close the hand can be uncomfortable. Class A is not suitable to be used, however Class B and C reached their full minimum and maximum value quite easily without struggle showing that they are working properly during full flexion and extension.

As seen in Fig. 4 the test shows that the sensor works by using the previous tightness value on people with varying forearm, so no customization is required for different people as it works on all the volunteers. The figure shows that all the test volunteer has about the same range of minimum and maximum value; give and take 30 value differences. What matters is how tight and secured the sensor is on the person. It also shows that the tighter the sensor is strapped on the forearm, the better it receives the IR signal as seen in Class B and Class C where there is a greater difference of maximum and minimum value.

This is because it is easier to detect the pressure. However, it can be uncomfortable when it is too tight for too long. When it is too loose, the sensor is still able to detect pressure, but it is difficult to reach the minimum value, that is when it is fully flexed. The only way to fully close the prosthetic hand is by fully flexing the forearm muscle up the point of exhaustion as seen in Fig. 3 graph.

The conclusion from this test is that Class A (Not Tight), everyone generally agreed that they were not able to flex the prosthetic hand successfully with the IR sensor receiving the signals very poorly. While Class C (Very Tight) had the best reception, it was uncomfortable for most of the volunteers. Class B (Tight) was tight enough while also receiving good signals from the IR sensor. When the sensor is securely tightened on the forearm and is not loose, only then it is able to reach its full value range making it easy to close and open the hand. So, Class B with tightness value of 11.28N is the most suitable classification to use out of all them.

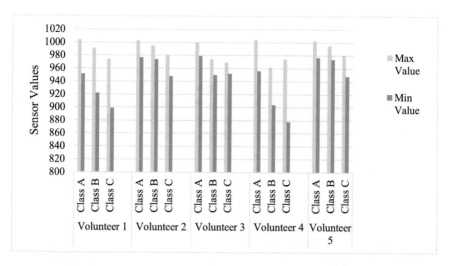

Fig. 4. Differences of minimum and maximum sensor values for 5 volunteers in the 3 Class.

3.3   Grasping Test

As seen in Fig. 5, the shape of the object being used to test the prosthetic hand with, corresponds to different types of grasp needed to hold it. The first picture shows the hand holding a cylinder which corresponds to a power grasp, the second and third picture is holding the rectangular object using tip grasp and lateral grasp and finally the last picture showing a spherical grasp by holding a spherical object.

The results after the test were almost the same as Kate et al. mentioned, which are the achievable grasp of power, precision and lateral where all the volunteers were able to do the grasp. However, we found out that it can also do the tip grasp. The results as

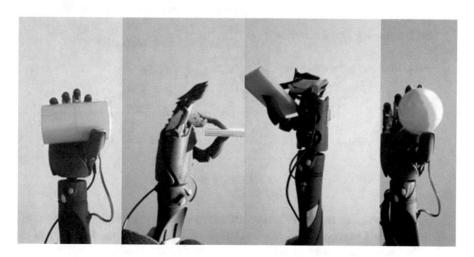

Fig. 5.   Testing different Grasps: Power, Tip, Lateral and Spherical.

seen in Fig. 5, depending on how the objects are shaped, the hand must grasp it in a different way. Precision grasp is similar to spherical grasp, for the purpose of this paper we will be using the term spherical grasp.

One of the main issues that was encountered during the test was that the hand did not have any friction. The problem made it especially difficult to do a precision grip with the ball. Although successful it was done with great difficulty. Another problem is that the way the hand is designed, it is unable to achieve grasp such as extension, and tripod as it is too rigid and not flexible enough.

3.4   ADL Test

This type of test allows us to see whether the hand would be functional in the Activities of Daily Living (ADL). For this test, we have set up different items that anyone uses daily such as water bottle, books and clothes, this can be seen in Fig. 6. The type of tightness used for this test is Class B as seen in Sect. 3.2. We have given the volunteer 1 min to complete a task depending on the item and to rate the difficulty from 1 to 10 with 1 being easy and 10 being impossible to do.

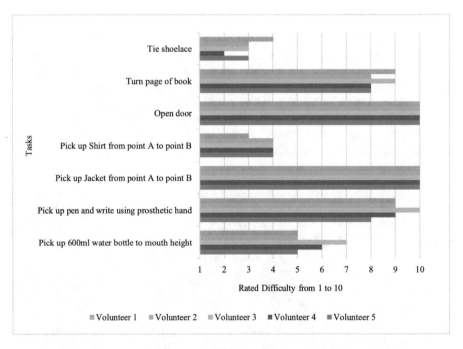

Fig. 6. ADL Test showing the volunteer's ability to complete a given task

Figure 6 shows the rate of difficulty for each volunteer, from 1 to 10. The result from the ADL test as seen in Fig. 6 is that the prosthetic hand can do 4 out of 7 of the ADL. The same issue came up with the grasp, is that the hand and fingers had no friction. This made it difficult to do tasks with objects that had no specific handles such as the turning pages of a book. Another task that we wanted the test volunteers to do was to write using the prosthetic hand. This was to see how well someone can write using the arm and the result was that the amputee was able to but not without difficulty. They had to find a proper angle, so the pen would stay in place by using the Lateral grasp. Lateral grasp was used in place of tripod due to the inability of the prosthetic arm to perform that gasp. The grasp was also used to tie shoelace with little difficulty. The prosthetic hand is unable to carry or pull heavy objects. This can be seen when trying to pull the door knob as well as trying to lift a jacket. The hand is only able to carry no more than 2 kilograms of weight. When doing so, the arm is in danger of breaking either from the servo motor or the printed parts. This makes its function limited for ADL as we do need to carry heavy weight at times.

4 Results and Discussion

4.1 Evaluation

The test results show that Class B of the tightness works best on the volunteers, which has 11.28N of tension after doing the Calibration test. While Class C worked responded better, however it was too tight on the volunteers making it very uncomfortable if used for a long time. The test also confirmed in Sect. 3.3 that the arm was able to have the grasp ability; Precision/Spherical, Power, Tip and Lateral grasp when doing the grasping test. The ADL test also confirms that the hand was able to complete 4 out of 7 tasks. We have also identified a few problems with the hand prosthetic.

One of the problems identified with the hand is the texture of it. As mentioned in the Grasping and ADL test is that the hand is too slippery and rigid. The filament that was used in the 3D-printer is PLA, it does not have the texture of a human skin to properly hold an object. This made it difficult to do certain task in the ADL test such as picking up water bottle and turning a page. The suggested improvement for this is to use a silicon cover on the fingers and palm to have a better grasp or to print the hand using a 3D-printer that supports a silicon like filament. This can be seen in Sayuk's paper [12] where he improved the fingers by adding an acetone smoothing to the surface and a rigid base for the palm. Another problem is that due to the design of the hand, it is difficult to achieve different type of grasp and the design only allows vertical movement of the fingers and not horizontal. This can be due to the IR sensor not being able to detect each finger movements as it only measures the flexing and extension. This can be improved by using additional sensor that can detect finger movements as well as using a servo motor that can move the fingers vertical and horizontally.

4.2 Cost

Traditionally, a customize prosthetic would cost around £10,000 average. This is considerably expensive and many of whom would not be able to afford. However, with the development of 3D-printed prosthetic open source project, a prosthetic can be made less than £400. Knowing the cost to make HACKberry is one of our aim as to know whether it is affordable or not. 3D Printer cost nothing as we already have a 3D-printer.

The only thing that had to be taken into consideration is the filament cost, which is about £37 for a large one. To print all the parts for HACKberry, we only needed one large PLA filament. Without a 3D printer, an average 3D-printer cost about £400 [13] or £310 to buy. The hardware parts cost about £300 bought from Japan including shipping cost. It is even cheaper if you can find the parts locally. The total for all the parts including a 3D-printer would be £647 and £337 provided you already own a 3D printer. The total breakdown of the cost can be seen in Table 1.

Table 1. Total cost to 3D-Print and Assemble the HACKberry designed prosthetic hand

Item	Cost (GBP)	Total price (GBP)
3D-Printer	£310	
PLA filament (large)	£37	
Hardware parts	£300	
	With 3D printer	£647
	Without 3D printer	£337

5   Conclusion and Future Work

Developing a low-cost 3D-printed prosthetic that is functional is possible for anyone who has an interest in it. However, it comes with a huge learning curve such as leaning to use a 3D-printer and using tools such as soldering. The total cost to create the HACKberry arm was around £337, as we already had a 3D printer. This is considerably cheaper than a customize made prosthetic. Cosmetic prosthesis from Bioengineering Institute Center for Neuroprosthetics, at the Worcester Polytechnic Institute can cost around $3000 USD to $5000 USD, while a functional one can go up to $20,000 USD to $30,000 USD or more [14].

The test for the HACKberry also shows that the prosthetic can work on any arm with different sizes provided that the sensor is strapped tight enough. This can be seen in the calibration test at Sect. 3.2. The functionality of the hand is however still limited. As seen from Sects. 3.3 and 3.4, the hand still does not have many different grasp a human hand does. This limits its ability to grasp certain item needed in our daily lives. This can be seen in Sect. 3.4 during the ADL testing where it was not able to grasp some of the item used for testing. The arm surface is plastic and rigid, this makes it difficult for it to grasp item due to the slippage caused by it. This issue occured in the test for both grasping test and ADL test. The other issue that the arm has is that it cannot lift heavy item limiting it to doing lightweight work.

Despite having this issue, the hand still proved to be functional at a low cost. It was able to do some of the ADL task with ease such as picking up a pen and turning a page of a book. The 3D-printed parts shows us the design possibility in the future to fix the issues mentioned above. The wide range of hardware that can be used can also improve the arm greatly. The design and creation of HACKberry proves to be a success at making a low-cost robotic prosthetic hand that is functional.

To improve the HACKberry arm, there are several potential possibilities that could be introduced. One of them is to include an additional EMG sensor to control the fingers seperately. This can be examplified by Nico Huchet with his version of HACKberry [9] using EMG instead of IR. This worked for him as he was already used to using EMG sensors for his prosthetic. Both sensors are available on SparksFun [15] with EMG under £30 The other improvement that can be made is by adding a silicon like texture to the HACKberry hand. This should increases its ability to grasp, This is due to the recurring issue as explained in Sects. 3.2 and 3.3.

References

1. Campbell, T., Williams, C., Ivanova, O., Garrett, B.: Could 3D printing change the world. In: Technologies, Potential, and Implications of Additive Manufacturing. Atlantic Council, Washington, DC (2011)
2. Doubrovski, Z., Verlinden, J.: Optimal design for additive manufacturing: opportunities and challenges. In: ASME 2011 (2011)
3. ten Kate, J., Smit, G., Breedveld, P.: 3D-printed upper limb prostheses: a review. Disabil. Rehabil. Assist. Technol. 12(3), 300–314 (2017)
4. Pylatiuk, C., Schulz, S., Döderlein, L.: Results of an Internet survey of myoelectric prosthetic hand users. Prosthet. Orthot. Int. 31(4), 362–370 (2007). https://doi.org/10.1080/03093640601061265
5. Zhou, J., Herscovici, D., Chen, C.: Parametric process optimization to improve the accuracy of rapid prototyped stereolithography parts. Int. J. Mach. Tools Manuf. 40(3): 363–379 (2000)
6. Exiii-hackberry.com. HACKberry |3D-printable open-source bionic arm (2018). http://exiii-hackberry.com/. Accessed 6 June 2018
7. Arduino.cc.: Arduino – Home (2018). https://www.arduino.cc/. Accessed 6 June 2018
8. Open Bionics: Open Bionics - turning disabilities into superpowers (2018). https://openbionics.com. Accessed 6 June 2018
9. My Human Kit: Exiii HACKberry myohand - My Human Kit (2018). http://myhumankit.org/en/tutoriels/myoelectric-exiii-hand/. Accessed 7 June 2018
10. ten Kate, J., Smit, G., Breedveld, P.: 3D-printed upper limb prostheses: a review. Disabil. Rehabil. Assist. Technol. 12(3), 300–314 (2017). https://doi.org/10.1080/17483107.2016.1253117
11. Merrett, G.: SHAP: Southampton Hand Assessment Procedure. Shap.ecs.soton.ac.uk (2018). http://www.shap.ecs.soton.ac.uk/. Accessed 7 June 2018
12. Sayuk, A.: Design and implementation of a low cost hand for prosthetic applications (2015)
13. 3D Printer Price: How Much Does a 3D Printer Cost? - 3D Insider. In: 3D Insider. http://3dinsider.com/cost-of-3d-printer/. Accessed 13 June 2018
14. McGimpsey, G,C., Bradford, T.: Nist.gov (2018). https://www.nist.gov/sites/default/files/documents/2017/04/28/239_limb_prosthetics_services_devices.pdf. Accessed 23 June 2018
15. SparkFun Electronics. Sparkfun.com. https://www.sparkfun.com/. Accessed 25 June 2018
16. Exiii-hackberry.com. http://exiii-hackberry.com/dw/doku.php?id=how_to_assemble:hand

Self-Trackam

Shiqah Natasya binti Muhammad Hadi[✉] and Mariatul Kiptiah binti Ariffin

School of Computing and Informatics, Universiti Teknologi Brunei, Gadong BE1410, Brunei
shiqah@outlook.com, mariatulit1990@gmail.com

Abstract. Leveraging face detection and tracking technology to enhance selfie experience on smart-phones, we design and implement the Self-Trackam system integrating various open-sources packages: Arduino, Processing 3, and Ketai library. Self-Trackam analyses video frames captured in real time to localize human faces in each frame. The system processes this information to position the phone which is mounted on two servos capable of panning horizontally and tilting vertically. The region of interest is automatically positioned in the centre of the frame using controls derived from faces positions. In this paper, the design and implementation details were discussed. Field test was conducted and results show that the Self-Trackam successfully enhances selfie-taking experiences.

Keywords: Computer vision · Face detection · Camera tracking · Arduino

1 Introduction

Many camera manufacturers provide delay timer so that the camera and the desired frame composition can be set before a photo is captured. This provides a fixed position and angle which is functional but lacks flexibility. Standard camera is capable of auto focusing but does not have any tracking feature, thus the camera cannot automatically track a moving *region of interest* (ROI) which may move out of the frame. As a result, a human operator is required to move the camera around the environment. This limitation could be addressed using the selfie stick, however, selfie stick extended over 1.5 m is not a practical application, and the stick would be visible within the frame. With access to technology and gadgets available in today's age, this constraint can be addressed.

In this work, the integration of computer vision and wireless communication technology enables autonomous detection and tracking of desired ROI. The program will detect an object of interest and positions the camera tracking hardware to maintain the ROI in the frame. This enables unmanned and dynamic coverage which could enhance photography experiences via autonomous control by tracking.

This work contributes in enhancing photography experiences by combining functions such as face detection, and tracking to move the camera in all axes for autonomous and manual control, into a single innovative product. All of which can be achieved using affordable and accessible components that will be discussed in Table 1.

The rest of this paper is organized as follows: the literature review describing related papers is covered in Sect. 2; Sect. 3 elaborates the implementation of the Self-Trackam;

S. Omar et al. (Eds.): CIIS 2018, AISC 888, pp. 224–233, 2019.
https://doi.org/10.1007/978-3-030-03302-6_20

Sect. 4 describes the testing conducted and results obtained; Conclusion is discussed in Sect. 5.

2 Literature Review

A study to determine the feasibility of a simplified face recognition machine by Bledsoe [1] is one of the pioneers in facial recognition research. The study aimed to evaluate whether it is possible to apply existing technology of that time to create a machine which can reliably solve simplified face recognition problem relying only pictures from a simple view to identify the identity of a person from a new unseen picture. The paper discusses various approaches to tackle variations introduced by translation, rotation, and occlusion. This study opened up research opportunity in the area of study.

The approach in [1] is holistic and considers all pixels in an image. This produces an image of size 300×300 pixels; it creates a vector of size 90,000. This is computational expensive. A low-dimensional procedure for the characterization of human faces is a paper presented by Sirovich and Kirby [2]. It implements the *Principle Component Analysis* (PCA) approach which was invented by Pearson [3] that maps the original data in a higher dimensional space to a relatively low-dimensional space within an acceptable error bound. The PCA presented in this study contributed to other facial recognition techniques such as the eigenfaces.

Representing face using all pixels in an image neglects structural information of facial components such as eyes, nose and mouth. Facial components could be effective descriptors to detect a face in an image. Although facial components can be computed from pixel intensity, the process is complex and expensive. In [4] the authors employ *Genetic Algorithm* (GA) to detect facial region with a complex background. GA detects the edges from an input image where it searches ellipse regions using GA based on the idea that human faces can be approximated by ellipsoid. The region is then judged whether it is a human face or not, confirming the validity and effectiveness of the method by computer simulation.

In [5], the authors propose a technique for direct visual matching of images using probabilistic measure of similarity based primarily on Bayesian, a *maximum a posteriori probability* (MAP) estimate of image intensity difference. The method is an instance of a non-Euclidean similarity measure that can be viewed as a general non-linear version of *Linear Discriminant Analysis* (LDA) making this method appealing in terms of computational simplicity and ease of implementation. When compared with the PCA, the LDA works better with large datasets having multiple classes, which is an important factor while reducing dimensionality.

2.1 Face Detection by the Haar Cascade Method

Haar cascade combines Haar-like features with the cascade process for face detection. A Haar-like feature considers adjacent rectangular regions at a specific location in a detection window, sums up the pixel intensities in each region and calculates the difference between these sums [6]. However it is an expensive method, hence the integral image has been devised to speed up the computation. This detection method was proposed by Viola and Jones [7] and while it was motivated primarily by the problem of face detection, it can be trained to detect a variety of object classes. The Haar cascade algorithm has three stages: (i) compute Haar features, (ii) Adaboost training, and (iii) Cascading classifiers [8].

Compute Haar Features: Haar feature are obtained by summing image pixel values within the rectangular area. The feature is the sum of the pixels within one rectangle subtracted from the sum of the pixels within another rectangle. This computation can be made very efficient with the use of the integral image concept. The integral image is the summed area table computed from the image, using the fact that the value in the summed area table.

$$I(x, y) = i(x, y) + I(x - 1, y) + I(x, y - 1) - I(x - 1, y - 1) \tag{1}$$

We can efficiently compute pixel information under a rectangular area.

Face detection with Haar cascade uses rectangle areas as features because all human faces share similar properties; they all have eyes, noses, forehead, chin and mouth. These regularities can be used to define the general structure of the face and matched to those Haar features. The face detection using Haar feature operates much faster than a pixel-based system.

Adaboost: *Adaboost* (Adaptive Boosting) is a machine learning meta-algorithm that can be used with many other types of learning algorithms to improve performance [9]. The Haar classifier cascade uses positive and negative images as training so it can detect human facial features. This algorithm constructs a strong classifier as a linear combination of weighted simple weak classifier.

$$h(x) = \text{sgn}\left(\sum_{j=1}^{M} \alpha_j h_j(x)\right) \tag{2}$$

The weak learning algorithm is designed to select the single rectangle feature which best separates the positive and negative examples For each feature, the weak learner determines the optimal threshold classification function, such that the minimum number of examples are misclassified. A weak classifier is a threshold function based on the feature f_j.

$$h_j(x) = \begin{cases} -s_j & \text{if } f_i < \theta_j \\ s_j & \text{otherwise} \end{cases} \tag{3}$$

The threshold value $\emptyset_j$ and the polarity $S_j \in \pm 1$ are determined in the training, as well as the coefficients α_j.

The Adaboost algorithm for classifier learning is shown below.

1. Given set of N positive and negative images (x^i, y^i). If image i is a face $y^i = 1$, else $y^i = -1$.

2. Initialize weights $W_{1,i} = \dfrac{1}{N}$ to each image i.
3. For each feature f_j with $j = 1,\ldots, M$
   i. Normalize the weights such that they sum to one.
   ii. Apply the feature to each image in the training set, and then find the optimal threshold and polarity $\varnothing_j$, S_j that minimizes the weighted classification error. That is

$$\theta_j,\ s_j = \arg\ \min_{\theta,s}\ \sum_{i=1}^{N} w_j^i \varepsilon_j^i \tag{4}$$

   Where

$$\varepsilon_j^i = \begin{cases} 0 \text{ if } y^i = h_j(x^i,\ \theta_j,\ s_j) \\ 1 \text{ otherwise} \end{cases} \tag{5}$$

   iii. Assign a weight α_j to h_j that is inversely proportional to the error rate.
   iv. Update the weights: w_{j+1}^i where it is reduced for the images I that were correctly classified.
4. The final strong classifier is completed using Eq. (2).

Cascading Classifier: The Cascade technique is employed in the Haar cascade; it is a degenerate decision tree which is the overall form of the detection process. Each classifier is a simple function made up of rectangular sums followed by a threshold. Higher thresholds yield classifiers with fewer false positives and a lower detection rate. Lower thresholds yield classifiers with more false positives and a higher detection rate.

The stages in the cascade are constructed by training classifiers using Adaboost and then adjusting the threshold to minimize false negatives. Each stage is trained by adding features until the overall desired accuracy and computation time is reached.

2.2 Related Works/Products

Face detection and tracking has been intensively studied in the past forty years. Haar cascade classifier has emerged and established as one of the best face detection algorithm. In our domain of interest, apart from the face detection, it is also possible to further extend the detection capability, for examples cloths and other accessories such as handbags, glasses. This could enable a precise tracking of specific targets without the need of a pre-training stage. Recent work by [10] shows that by combing local features from SIFT [11] or SURF [12], and global feature from Camshift [13], objects can be anonymously tracked.

Self-Trackam make use of detection and tracking functions, similar products that exist and utilizes those features as a marketable product are the selfie drone, soloshot3, and pixy.

Soloshot3 comes with a robotic base, wearable tag and a custom designed camera. The tag is to be attached to the subject while the robotic base detects and tracks the tag automatically [14]. The Mavic Pro drone has an ActiveTrack functionality where user can command the drone to track in ways such as behind, in front, alongside the subject, or trained on the subject while the drone fly almost everywhere [15]. Pixy is a vision system that uses a colour-based filtering algorithm to detect objects, which is robust when it comes to lightings and exposure changes. It can store up to 7 different colour signatures which can be used to identify or detect hundreds of objects at the same time [16].

While the Self-Trackam, soloshot3 and the Mavic Pro are similar for its use in photography or filming, the Pixy is the only device that uses a vision sensor, primarily used for DIY robotics.

3   Implementation of Self-Trackam

The software and library used to develop Self-Trackam application are Processing 3 [17]; a Java based programming language, and Ketai library [18] that is imported to compute the face detection algorithm. The Arduino IDE [19] is coded in C# that processes the tracking aspect of the system. Hardware components listed in Table 1 were assembled; servos are attached to the phone holder to prepare the mount, and placed on a tripod with adjustable height.

Table 1. Function description of each hardware components.

Hardware	Function description
Android Smartphone	The vision sensor is the camera of the Android smart phone, where it will analyse the video frames captured and execute the face detection program which is the Self-Trackam application
Bluetooth Module	The Bluetooth module transmits the coordinates and position of the face detected from the android smart phone to the Arduino
Arduino Uno	Receives the coordinates from either the Bluetooth module or IR receiver, and controls the rotation of the servos to pan and tilt
Servo	Pan and tilt hardware
IR Remote Control	To toggle between autonomous tracking and manual control of the pan, tilt and capture
IR Receiver	Receives commands from the remote control and relays to the Arduino

3.1   Features of Self-Trackam

The main features of the Self-Trackam are the face detection and tracking capability. Wireless communications are provided by the Bluetooth and infrared modules. Bluetooth is used as communication link between the smart-phone and Arduino, while the infrared allowed operators to remotely control the Arduino. The relationship and the flow of data between the components are illustrated in Fig. 1.

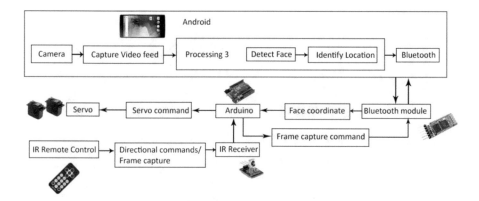

Fig. 1. Architecture design and functionality block of Self-Trackam

Manual control and autonomous control can be toggled using the remote control; users can control the pans and tilts using the directional buttons. The remote control can calibrate the servo back to centre, as well as capturing photos which are stored in the phone's internal memory. A shutter sound effect will be played by the system to notify the user that a photo is successfully taken and saved. This uses the cassette library by Shlomihod [20], a processing audio implementation for android.

Potential Features: Availability and accessibility to technology gives the Self-Trackam potential enhancement for future upgrade such as facial recognition capabilities. This allows the system to process captured facial images and perform comparison algorithm against an existing database for a more accurate tracking to perform as the irrelevant faces would be ignored. However facial recognition requires advance programming that is too difficult to implement in the current state of the system.

We could also enhance the frame resolution and refresh rates that allow high definition video feed in real-time without compromising face detection performance. Another feature we would like to add is to introduce camera toggle functionality that allows the video feed to switch between the back and frontal camera from the application.

4   Testing and Results

An experiment was conducted to ensure that the functionality and features of Self-Trackam were operational, by primarily assessing its reliability and consistency of the system's face detection and motion tracking functionality.

Test for the prototype was set up by placing a fixed face image in a 3D space. The test demonstrated the tracking hardware adjusting its pan and tilt to ensure the face detected was centred in the frame. As seen in the top images in Fig. 2, the first image shows the face is on the right side of the frame and the second image displays the image centred by the servos.

Fig. 2. Self-Trackam's detection and tracking of face image.

The system is able to detect multiple faces in a frame. However, it has trouble detecting when too many faces are presented as shown in the bottom images of Fig. 2. The limitation with the low resolution capture causes faces to become less detectable due to obscured facial features or increased in distance between the camera and subject.

Three various face type are used to experiment the detection rate of the application to obtain its maximum distance and response rate. It is observed to see whether different facial feature can affect the performance of the system.

Images of face data were printed in equal sizes and placed side by side. As presented in Table 2, face data 1 could be detected from furthest away, with small difference with face data 3, while face data 2 had to be closest. Distance values are estimated as lightings and orientation affected the results, as it varies at times. When all faces are within range, face data 1 and 3 were detected simultaneously and almost immediate. At times face data 2 was also detected at the same time, however more often it was couple of seconds behind.

Table 2. Various face data detection rate.

Face data	Maximum distance (centimeters)	Response rate (seconds)
Data 1	90	2
Data 2	75	2–4
Data 3	85	2

This shows that different face type, lightings, orientation, and background contribute toward the performance of the face detection.

4.1   Beta-Testing

Testing was conducted among the students of Universiti Teknologi Brunei. Before operating the Self-Trackam, the volunteers were briefed in the mechanics and functionality of the application.

The Self-Trackam successfully detected the faces of the volunteers and followed where they moved, provided the face remained within the frame. The system suffered lag which is believed to be caused by hardware limitation. It is possible the 4 GB RAM size of the android phone is not sufficient to process detection algorithm. The volunteers had to look toward the camera directly and stay within distance; else the face won't be detected due to the low resolution.

When multiple faces were detected in the frame, the camera could not stabilize to a specific location as it attempts to centre all the faces. However the outcome was still satisfactory as the camera managed to keep all the faces within the frame.

The volunteers were able to capture the photos instantly from a distance via the remote control, and the shutter sound effect is played when a photo is successfully saved. While the real-time video feed have a low resolution, the saved photo has a resolution of 1280×720.

4.2   Feedbacks

Feedbacks were analysed from participant's responses to support the results, in a form of questionnaires taken after conducting the test or watching the demonstration. There are 35 participants; 19 males and 16 females who are aged 20 years old and above.

Most of the participants were not familiar with Computer Vision. Only 14% of the participants had interest in taking selfies. However majority agreed with the problem and issues that came with taking the perfect selfie that the Self-Trackam hopes to overcome.

As shown in Fig. 3, majority of participants agreed that the face detection and tracking capability would ease the use of taking, and the manual remote control feature was proven to be convenient. Overall, they also believed that the technology and features implemented in Self-Trackam would likely help enhance photography and selfie taking experience. While 55% believed the product is marketable in Brunei were likely, 17% were opposite and the rest were neutral.

The Self-Trackam is not without limitation and drawbacks. For instance, the system lacks visible live preview as user cannot see what is displayed on the phone's LCD when using the back camera. As opposed to Bluetooth, infrared is an aging technology, and devices need to be directly pointed toward the IR receiver to work. Thus, any barrier could block the infrared communication between the remote and receiver module. There is also room for improvement for the low quality video feed as the current frame is limited to 320×240 resolution.

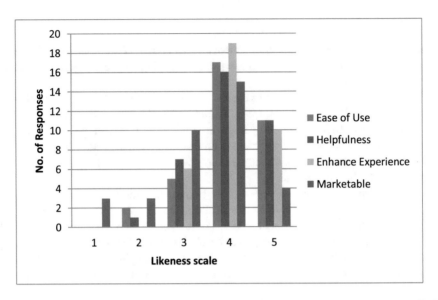

Fig. 3. Data analysed from the questionnaires responded by participants, where 1 is very unlikely and 5 is very likely.

5 Conclusion

Self-Trackam is an Arduino-Android based application capable of face detection, camera tracking, and uses wireless communication to enhance selfies and photography experience.

Self-Trackam involves the process of detecting and tracking the user's face, ensuring the face is framed in the centre. Using the Android mobile phone camera, mounted on two servos; to pan and tilt enabling it to rotate horizontally and vertically. Real-time video frames are analysed to detect facial features, and if the target object moves, the program will command the servo to also move so that the face is always in the scene.

The face detection algorithm is processed with the Android application in the smart phone where it sends servo commands to the Arduino via Bluetooth to be tracked. Infrared communication enables wireless connectivity for manual control of the camera, so user can operate the pan and tilt, and frame capture with the press of the buttons from distance.

During my research on the project, I have discovered the effectiveness of computer vision on face detection which relates to the problem of this project while also broadening my knowledge in studying the area of computer vision technique, in Haar Cascade.

Through the testing of the application with the participants and the questionnaires, the results supported by the feedbacks analysed shows that through the features offered in Self-Trackam, the aims and objective of the project were met; it can help enhance photography and selfie-taking experiences. Despite the limitations, the results obtained are promising.

Acknowledgments. I am honoured and humbled that I have been given the opportunity to work with my supervisor, Associate Professor Somnuk Phon-Amnuaisuk who has tirelessly supported me with his endless guidance and mentoring through the course of my study in Universiti Teknologi Brunei. I would also like to thank my peers and those important to me, and lastly, to the Brunei Government's scholarship for the opportunity to attend UTB and pursue my studies.

References

1. Bledsoe, W.W.: A proposal for a Study to Determine the Feasibility of a Simplified Face Recognition Machine. Panoromic Research, Inc. (1963)
2. Sirovich, L., Kirby, M.: Low-dimensional procedure for the characterization of human faces. Josa a **4**(3), 519–524 (1987)
3. Pearson, K.: On Lines and Planes of closes fit to systems of points in space. Philos. Mag. **2**(11), 559–572 (1901). Series 6
4. Yokoo, Y., Hagiwara, M.: Human face detection method using genetic algorithm. IEEJ Trans. Electron. Inf. Syst. **117**(9), 1245–1252 (1997)
5. Moghaddam, B., Jebara, T., Pentland, A.: Bayesian face recognition. Pattern Recogn. **33**(11), 1771–1782 (2000)
6. Crow, F.: Summed-area tables for texture mapping. In: Proceedings of SIGGRAPH, vol. 18, no. 3, pp. 207–212 (1984)
7. Viola, P., Jones, M.: Rapid object detection using a boosted cascade of simple features. In: Proceedings of the 2001 IEEE Computer Society Conference on Computer Vision and Pattern Recognition, CVPR 2001, vol. 1, p. I. IEEE (2001)
8. "Viola–Jones object detection framework" in Wikipedia: The Free Encylopedia. https://en.wikipedia.org/wiki/Viola–Jones_object_detection_framework. Accessed 27 Mar 2017
9. Freund, Y., Schapire, R.E.: A decision-theoretic generalization of on-line learning and an application to boosting. J. Comput. Syst. Sci. **55**(1), 119–139 (1997)
10. Phon-Amnuaisuk, S., Ahmad, A.: Tracking and identifying a changing appearance target. In: Proceedings of the 9th International Multi-disciplinary Trends in Artificial Intelligence Workshop (MIWAI 2015), Fuzhou, pp. 245–252 (2015)
11. Lowe, D.G.: Object recognition from local scale-invariant features. In: Proceedings of the Seventh IEEE International Conference on Computer Vision, Kerkyra, vol. 2, pp. 1150–1157 (1999)
12. Bay, H., Ess, A., Tuytelaars, T., Van Gool, L.: SURF: speeded up robust features. Comput. Vis. Image Underst. (CVIU) **110**(3), 346–359 (2008)
13. Bradski, G.R.: Computer vision face tracking for use in a perceptual user interface (1998)
14. Soloshot Homepage. https://soloshot.com/. Accessed 18 Aug 2018
15. DJI Mavic Pro. http://store.dji.com/product/mavic-pro?gclid=EAIaIQobChMIpvew5PSy1gIVlgMqCh0QfgNaEAAYASAAEgKwH_D_BwE. Accessed 18 Aug 2018
16. CMUcam Homepage. http://cmucam.org/projects/cmucam5. Accessed 18 Aug 2018
17. Processing Homepage. https://processing.org/. Accessed 18 Aug 2018
18. Ketai Homepage. http://ketai.org. Accessed 18 Aug 2018
19. Arduino Homepage. https://www.arduino.cc/en/Main/Software. Accessed 19 Aug 2018
20. Github, Shlomihod, Cassette. https://github.com/shlomihod/cassette. Accessed 4 Apr 2018

Author Index

© Springer Nature Switzerland AG 2019
S. Omar et al. (Eds.): CIIS 2018, AISC 888, pp. 235–236, 2019.
https://doi.org/10.1007/978-3-030-03302-6

Printed in the United States
By Bookmasters